LONDON MATHEMATICAL SOCIETY STUDENT TEXTS

47 Introductory lectures on rings and modules. JOHN A. BEACHY
48 Set theory, ANDRÁS HAJNAL & PETER HAMBURGER. Translated by ATTILA MATE
49 An introduction to K-theory for C*-algebras, M. RØRDAM, F. LARSEN & N. J. LAUSTSEN
50 A brief guide to algebraic number theory, H. P. F. SWINNERTON-DYER
51 Steps in commutative algebra: Second edition, R. Y. SHARP
52 Finite Markov chains and algorithmic applications, OLLE HÄGGSTRÖM
53 The prime number theorem, G. J. O. JAMESON
54 Topics in graph automorphisms and reconstruction, JOSEF LAURI & RAFFAELE SCAPELLATO
55 Elementary number theory, group theory and Ramanujan graphs, GIULIANA DAVIDOFF, PETER SARNAK & ALAIN VALETTE
56 Logic, induction and sets, THOMAS FORSTER
57 Introduction to Banach algebras, operators and harmonic analysis, GARTH DALES *et al*
58 Computational algebraic geometry, HAL SCHENCK
59 Frobenius algebras and 2-D topological quantum field theories, JOACHIM KOCK
60 Linear operators and linear systems, JONATHAN R. PARTINGTON
61 An introduction to noncommutative Noetherian rings: Second edition, K. R. GOODEARL & R. B. WARFIELD, JR
62 Topics from one-dimensional dynamics, KAREN M. BRUCKS & HENK BRUIN
63 Singular points of plane curves, C. T. C. WALL
64 A short course on Banach space theory, N. L. CAROTHERS
65 Elements of the representation theory of associative algebras I, IBRAHIM ASSEM, DANIEL SIMSON & ANDRZEJ SKOWROŃSKI
66 An introduction to sieve methods and their applications, ALINA CARMEN COJOCARU & M. RAM MURTY
67 Elliptic functions, J. V. ARMITAGE & W. F. EBERLEIN
68 Hyperbolic geometry from a local viewpoint, LINDA KEEN & NIKOLA LAKIC
69 Lectures on Kähler geometry, ANDREI MOROIANU
70 Dependence logic, JOUKU VÄÄNÄNEN
71 Elements of the representation theory of associative algebras II, DANIEL SIMSON & ANDRZEJ SKOWROŃSKI
72 Elements of the representation theory of associative algebras III, DANIEL SIMSON & ANDRZEJ SKOWROŃSKI
73 Groups, graphs and trees, JOHN MEIER
74 Representation theorems in Hardy spaces, JAVAD MASHREGHI
75 An introduction to the theory of graph spectra, DRAGOŠ CVETKOVIĆ, PETER ROWLINSON & SLOBODAN SIMIĆ
76 Number theory in the spirit of Liouville, KENNETH S. WILLIAMS
77 Lectures on profinite topics in group theory, BENJAMIN KLOPSCH, NIKOLAY NIKOLOV & CHRISTOPHER VOLL
78 Clifford algebras: An introduction, D. J. H. GARLING
79 Introduction to compact Riemann surfaces and dessins d'enfants, ERNESTO GIRONDO & GABINO GONZÁLEZ-DIEZ
80 The Riemann hypothesis for function fields, MACHIEL VAN FRANKENHUIJSEN
81 Number theory, Fourier analysis and geometric discrepancy, GIANCARLO TRAVAGLINI
82 Finite geometry and combinatorial applications, SIMEON BALL
83 The geometry of celestial mechanics, HANSJÖRG GEIGES
84 Random graphs, geometry and asymptotic structure, MICHAEL KRIVELEVICH *et al*
85 Fourier analysis: Part I - Theory, ADRIAN CONSTANTIN
86 Dispersive partial differential equations, M. BURAK ERDOĞAN & NIKOLAOS TZIRAKIS

LONDON MATHEMATICAL SOCIETY STUDENT TEXTS

Managing Editor: Professor D. Benson,
Department of Mathematics, University of Aberdeen, UK

[illegible]

London Mathematical Society Student Texts 86

Dispersive Partial Differential Equations

Wellposedness and Applications

M. BURAK ERDOĞAN
University of Illinois, Urbana-Champaign

NIKOLAOS TZIRAKIS
University of Illinois, Urbana-Champaign

CAMBRIDGE
UNIVERSITY PRESS

University Printing House, Cambridge CB2 8BS, United Kingdom

One Liberty Plaza, 20th Floor, New York, NY 10006, USA

477 Williamstown Road, Port Melbourne, VIC 3207, Australia

4843/24, 2nd Floor, Ansari Road, Daryaganj, Delhi - 110002, India

79 Anson Road, #06-04/06, Singapore 079906

Cambridge University Press is part of the University of Cambridge.

It furthers the University's mission by disseminating knowledge in the pursuit of education, learning and research at the highest international levels of excellence.

www.cambridge.org
Information on this title: www.cambridge.org/9781107149045

First published 2016

A catalogue record for this publication is available from the British Library

ISBN 978-1-107-14904-5 Hardback

Dedicated to our Children

Eda, Sifis, and Sofia.

Contents

Preface

This book is intended for beginning graduate students in mathematics with some background in real and complex analysis who are interested in pursuing research in nonlinear dispersive partial differential equations (PDEs). This area has become exceedingly technical branching out into many different directions in recent decades. With this book, our aim is to provide a gentle introduction to the basic methods employed in this area in a self contained manner and in the setting of a few model equations. However, we should note that these methods are more generally applicable, and play a central role in modern research in nonlinear dispersive PDEs.

We designed this book having in mind a semester-long course in this area for advanced undergraduate and beginning graduate students. For that reason, we restricted the discussion to a few basic equations while providing complete details for each topic covered. We have also included many exercises that supplement and clarify the material that is discussed in the main text. After reading our book, a student should be able to read recent research papers in nonlinear dispersive PDEs and start making contributions.

There are several books, including Cazenave [28, 29, 30], Bourgain [20], Sulem–Sulem [138], Tao [143], and Linares–Ponce [105], which cover a large proportion of this area. In comparison, our book concentrates more on problems with periodic boundary conditions and aims to introduce the wellposedness techniques of model equations, such as the Korteweg de-Vries (KdV) and nonlinear Schrödinger (NLS) equations. The methods we describe also apply to various dispersive models and systems of dispersive equations, such as the fractional Schrödinger equation and the Zakharov system. In cases where the model equations are integrable, such as the periodic KdV and cubic NLS equations, alternative methods based on the symmetries and the structure of the equations have been developed. We refer the interested reader to Pöschel–Trubowitz [123], Kuksin [99], and Kappeler–Topalov [81] for complete integrability and inverse scattering techniques that extend some of the analytical results presented here. However, we should mention that we will not make use of any complete integrability methods in this book.

The KdV and NLS equations are the simplest models, combining the effects of dispersion and nonlinear interactions. The KdV equation describes very diverse physical phenomena, such as surface water waves in shallow water, propagation of ion-acoustic waves in cold plasma, and pressure waves in liquid-gas bubble mixture. In the case of shallow water, one normally does not work with the full water wave equation but uses approximate models to study the evolution. In particular, the KdV equation describes unidirectional small amplitude long waves on a fluid surface.

The NLS equation arises in a number of physical models in the theory of nonlinear optics. For example, it frequently appears as the leading approximation of the envelope dynamics of a quasi-monochromatic plane wave propagating in a weakly nonlinear dispersive medium. It also arises in the description of Bose–Einstein condensation. Another equation we consider in this book is the fractional NLS equation, which is a basic model in the theory of fractional quantum mechanics. It is also used as a model describing charge transport in bio polymers like DNA.

The NLS equation, having a power nonlinearity, is easier to deal with in high regularity spaces by Sobolev embedding techniques. For lower regularity solutions on $\mathbb{R}^n$, Strichartz estimates are the main tools to establish wellposedness. On the other hand, in the case of the KdV equation, the derivative nonlinearity makes the problem more complicated. In fact, even the existence of smooth solutions requires more elaborate techniques. The situation is even more complicated for initial value problems on bounded domains, where the dispersion is weaker, and the wellposedness is harder to establish, especially in low regularity spaces.

In recent decades, a variety of techniques utilizing harmonic analysis methods were applied in conjunction with classical PDE tools to address these difficulties. Most of these techniques rely on time averaging via space-time norm estimates. Along these lines, we discuss Strichartz estimates, which is a very efficient method of establishing the wellposedness of dispersive PDEs with power type nonlinearities. We also discuss oscillatory integral techniques, which is based on the representation of the solution using the Fourier transform in the space variable. This technique is very efficient when dealing with equations with derivative nonlinearities. In addition, we present the restricted norm method using an anisotropic space-time Sobolev norm which takes into account the distance between the space-time Fourier supports of the linear and nonlinear solutions.

We now give a short summary of the contents of the book, which is divided into five chapters. In the first chapter, we recall without proof basic results from analysis that will be used throughout the text. Although we expect the reader to be familiar with basic harmonic analysis techniques, all the results we need in this book are outlined in this chapter.

In the second chapter, we concentrate on linear dispersive equations on the real line and on the torus. The methods are perturbative around the linear solution and the mapping properties of the linear propagator are extremely important in studying nonlinear counterparts. In particular, to find out which space is suitable in order to analyze the nonlinear solution, one needs to understand the decay and smoothing properties of the linear solution. We thus establish Strichartz estimates, Kato smoothing, and maximal function estimates for equations on the real line, and Strichartz estimates for equations on the torus. In this book, we make an effort to present various applications of the methods we discuss which are not found in other books in the area. One such application is the so-called Talbot effect for nonlinear dispersive PDEs on the torus. We finish Chapter 2 with a discussion of the Talbot effect for linear equations. This discussion is also useful for understanding the differences between the dynamics of dispersive PDEs on bounded and unbounded domains.

In the third chapter, we study basic wellposedness methods for the KdV equation on the torus and the real line, and the NLS equation on the torus. We start with the energy method based on parabolic regularization and the conservations laws of the equation. This method applies equally well to dispersive and nondispersive evolution equations, and it is a useful tool for studying smooth solutions. Then we discuss the oscillatory integral method of Kenig–Ponce–Vega, which uses the dispersive estimates established in Chapter 2. This method is useful mainly for the equations on $\mathbb{R}^n$. We continue with the restricted norm method of Bourgain. We then proceed to establish a version of the normal form transform, which we use to establish nonlinear smoothing and unconditional wellposedness results. We close this chapter with a thorough discussion of illposedness results.

In the fourth chapter, we study rough data global wellposedness and nonlinear smoothing of model dispersive equations. In particular, we present Bourgain's high–low decomposition method to establish global solutions when no a priori bounds are available. We also discuss the method of almost conserved quantities of Colliander–Keel–Staffilani–Takaoka–Tao, which can be considered as a refinement of the high–low decomposition method.

Finally, in the fifth chapter we present some applications of the techniques we developed in the previous chapters. More precisely, we study the growth bounds for higher order Sobolev norms, almost everywhere convergence to initial data for rough nonlinear solutions, the Talbot effect for nonlinear equations, and the existence and regularity of global attractors for dissipative and dispersive equations.

During the writing of this book the first author was partially supported by NSF grants DMS-1201872 and DMS-1501041. The second author was partially supported by the NSF grant DMS-0901222, the Simons Foundation grant #355523, and the University of Illinois Research Board grant RB-14054.

M. Burak Erdoğan
Nikolaos Tzirakis

Urbana, Illinois
December 2015

Notation

$\langle x \rangle$	$= \sqrt{1+\lvert x\rvert^2}$, $x \in \mathbb{R}^n$.
$B(x,r)$	The ball centered at x with radius r.
$A \lesssim B$	$A \leq CB$, where $C > 0$ is an absolute constant.
$A \approx B$	$A \lesssim B$ and $B \lesssim A$.
$A \lesssim B^{s\pm}$	$A \lesssim B^{s\pm\epsilon}$ for any $\epsilon > 0$.
$A \ll B$	$A \leq \frac{1}{C}B$, where C is a sufficiently large constant.
$A = O(B)$	$A \lesssim B$.
$A = o(B)$	$\lim \frac{A}{B} = 0$.
$\mathbb{R}$	The field of real numbers.
$\mathbb{T}$	The torus $\mathbb{R}/2\pi\mathbb{Z}$.
$\mathbb{C}$	The field of complex numbers.
$L^p(K)$	The Lebesgue spaces of measurable functions (for $K = \mathbb{T}$ or $\mathbb{R}$): $\left\{f : K \to \mathbb{C} : \lVert f\rVert_{L^p}^p := \int_K \lvert f\rvert^p < \infty\right\}$, $p \in [1,\infty)$, with the usual modification when $p = \infty$.
ℓ^p	$= \left\{a : \mathbb{Z} \to \mathbb{C} : \lVert a\rVert_{\ell^p}^p = \sum_{k\in\mathbb{Z}} \lvert a_k\rvert^p < \infty\right\}$, $p \in [1,\infty)$.
$\langle f,g\rangle_{L^2(K)}$	$= \int_K \overline{f(x)}g(x)dx$, $K = \mathbb{R}$ or $\mathbb{T}$.
$\mathcal{F}$	Fourier transform on $\mathbb{R}$: $\mathcal{F}f(\xi) = \widehat{f}(\xi) = \frac{1}{\sqrt{2\pi}}\int_{\mathbb{R}} f(x)e^{-i\xi x}dx$, $\xi \in \mathbb{R}$, or Fourier series on $\mathbb{T}$: $\mathcal{F}f(k) = \widehat{f}(k) = \frac{1}{2\pi}\int_0^{2\pi} f(x)e^{-ikx}dx$, $k \in \mathbb{Z}$.
$\mathcal{F}^{-1}$	Inverse Fourier transform on $\mathbb{R}$: $\mathcal{F}^{-1}\widehat{f}(x) = \frac{1}{\sqrt{2\pi}}\int_{\mathbb{R}} \widehat{f}(\xi)e^{i\xi x}dx$, $x \in \mathbb{R}$, or on the torus: $\mathcal{F}^{-1}\widehat{f}(x) = \sum_{k\in\mathbb{Z}} \widehat{f}(k)e^{ikx}$.

$f * g$	The convolution of f and g.
D^s	The multiplier operator with the multiplier $\lvert\xi\rvert^s$, $s \in \mathbb{C}$.
J^s	The multiplier operator with the multiplier $\langle\xi\rangle^s$, $s \in \mathbb{C}$.
$\lVert f\rVert_{H^s}$	$= \lVert J^s f\rVert_{L^2}$, $s \in \mathbb{R}$.
$\lVert f\rVert_{\dot{H}^s}$	$= \lVert D^s f\rVert_{L^2}$, $s \geq 0$.
$C_t^0 H_x^s$	The Banach space of H^s valued continuous functions with the norm $\sup_t \lVert u(t,\cdot)\rVert_{H^s}$
$\lVert T\rVert_{X\to Y}$	The operator norm of a bounded linear operator $T : X \to Y$ between Banach spaces X and Y.
X'	The dual of a topological vector space X.
T^*	The adjoint of an operator T.
$C^\infty(K)$	$= \{f : K \to \mathbb{C} : f \text{ is infinitely differentiable}\}, \quad K = \mathbb{R} \text{ or } \mathbb{T}$.
$C_0^\infty(\mathbb{R})$	$= \{f \in C^\infty(\mathbb{R}) : f \text{ is compactly supported}\}$.
$P_{m,n}(f)$	$= \left\lVert \langle x\rangle^m f^{(n)}(x)\right\rVert_{L^\infty}$.
$\mathcal{S}(\mathbb{R})$	Schwartz space: $\{f \in C^\infty(\mathbb{R}) : P_{m,n}(f) < \infty, m, n \geq 0\}$.
$\mathcal{D}(\mathbb{R})$	$= (C_0^\infty(\mathbb{R}))'$, the space of distributions.
$\mathcal{D}(\mathbb{T})$	$= (C^\infty(\mathbb{T}))'$, the space of periodic distributions.
$\mathcal{S}'(\mathbb{R})$	The space of tempered distributions.
$u(\phi)$	The action of the distribution u on the test function ϕ.
H	The Hilbert transform: $Hf(x) = \mathcal{F}^{-1}\left(i\mathrm{sign}(\cdot)\widehat{f}(\cdot)\right)(x)$.
$P_k f$	The Littlewood–Paley projection on to the frequencies $\approx 2^k$. We define $P_{\leq k}$, $P_{\geq k}$ similarly.

$P_N f$	The Littlewood–Paley projection on to the frequencies $\approx N$.
$B^s_{p,\infty}$	The Besov space defined by the norm: $\|f\|_{B^s_{p,\infty}} := \sup_{j\geq 0} 2^{sj}\|P_j f\|_{L^p}$.
M	Hardy–Littlewood maximal function: $Mf(x) = \sup_{r>0} \frac{1}{\|B(x,r)\|} \int_{B(x,r)} \|f(y)\| \, dy$.
W_t	The propagator of the Airy equation, $W_t g = e^{-t\partial_{xxx}} g$.
W^γ_t	The propagator of the weakly damped Airy equation: $W^\gamma_t g = e^{-t\partial_{xxx} - t\gamma} g$.
$X^{s,b}$	The restricted norm space. In the case of the KdV equation, it is defined by the norm: $\|u\|_{X^{s,b}} = \left\| \widehat{u}(\tau, \xi) \langle \tau - \xi^3 \rangle^b \langle \xi \rangle^s \right\|_{L^2_{\tau,\xi}}$.

1

Preliminaries and tools

The material we present in this book relies heavily on basic harmonic analysis tools on the real line $\mathbb{R}$ and on the torus $\mathbb{T} = \mathbb{R}/2\pi\mathbb{Z}$. There are many excellent textbooks on the subject, e.g. Katznelson [84], Stein–Weiss [135], Stein [133], Folland [61], Wolff [154], and Muscalu–Schlag [116]. In this preliminary chapter, we state without proof the results we need in order to develop the wellposedness theory of dispersive partial differential equations (PDEs).

We first recall the Lebesgue spaces of measurable functions (for $K = \mathbb{T}$ or $\mathbb{R}$):

$$L^p(K) = \left\{ f : K \to \mathbb{C} : \|f\|_{L^p}^p := \int_K |f|^p < \infty \right\}, \quad p \in [1, \infty),$$

$$L^\infty(K) = \{ f : K \to \mathbb{C} : \|f\|_{L^\infty} := \operatorname{esssup}|f| < \infty \},$$

and Hölder's inequality

$$\left| \int_K fg \right| \le \|f\|_{L^p(K)} \|g\|_{L^q(K)}, \quad 1 \le p, q \le \infty, \quad \frac{1}{p} + \frac{1}{q} = 1.$$

Also recall

$$\ell^p = \left\{ a : \mathbb{Z} \to \mathbb{C} : \|a\|_{\ell^p}^p = \sum_{k \in \mathbb{Z}} |a_k|^p < \infty \right\}, \quad p \in [1, \infty), \quad \text{and}$$

$$\ell^\infty = \left\{ a : \mathbb{Z} \to \mathbb{C} : \|a\|_{\ell^\infty} = \sup_{k \in \mathbb{Z}} |a_k| < \infty \right\}.$$

For linear operators between L^p spaces, we have the Riesz–Thorin interpolation theorem, see Folland [61]:

Theorem 1.1 *Let T be a linear operator mapping $L^{p_0} + L^{p_1}$ to $L^{q_0} + L^{q_1}$. Fix*

$\theta \in (0,1)$ and define

$$\frac{1}{p_\theta} = \frac{1-\theta}{p_0} + \frac{\theta}{p_1}, \quad \frac{1}{q_\theta} = \frac{1-\theta}{q_0} + \frac{\theta}{q_1}.$$

Then T maps L^{p_θ} to L^{q_θ}, and

$$\|T\|_{L^{p_\theta}\to L^{q_\theta}} \le \|T\|_{L^{p_0}\to L^{q_0}}^{1-\theta} \|T\|_{L^{p_1}\to L^{q_1}}^{\theta}.$$

An easy consequence of the above theorem is Young's inequality for the convolution of two functions (see Exercise 1.1)

$$\|f * g\|_{L^r} \le \|f\|_{L^p}\|g\|_{L^q}, \quad 1 + \frac{1}{r} = \frac{1}{p} + \frac{1}{q}, \quad 1 \le p, q, r \le \infty. \tag{1.1}$$

The corresponding statement holds also for the ℓ^p spaces.

Another useful convolution inequality is the Hardy–Littlewood–Sobolev theorem; see Stein [133]:

Theorem 1.2 *For any $1 < p < r < \infty$*

$$\big\| |\cdot|^{-\alpha} * f \big\|_{L^r(\mathbb{R})} \lesssim \|f\|_{L^p(\mathbb{R})}, \quad \alpha = 1 + \frac{1}{r} - \frac{1}{p}.$$

A useful extension of Riesz–Thorin theorem is the complex interpolation theorem of Stein [132]:

Theorem 1.3 *Let $\{T_z\}$ be a family of linear operators analytic in the strip $\{z \in \mathbb{C} : 0 < Re(z) < 1\}$ and continuous on the closure. Namely, for any test functions f, g, the inner product $\langle f, T_z g\rangle$ is analytic on the strip and continuous on the closure. Assume that there exists $b < \pi$ so that for any simple functions f, g, and for any z in the strip*

$$|\langle f, T_z g\rangle| \le C_{f,g} e^{b|\Im(z)|}.$$

Also assume that

$$\|T_{0+iy}\|_{L^{p_0}\to L^{q_0}} \le M_0(y), \quad \|T_{1+iy}\|_{L^{p_1}\to L^{q_1}} \le M_1(y)$$

and that M_j, $j = 1, 2$, grow at most exponentially as $y \to \pm\infty$. Then for all $0 \le \theta \le 1$

$$\|T_\theta\|_{L^{p_\theta}\to L^{q_\theta}} \le C,$$

where C depends on M_j and θ.

For a Banach space X, the dual space X' is the space of all bounded linear maps from X to $\mathbb{C}$. From now on, we will use p' for the dual exponent $\frac{p}{p-1}$ of p, and similarly for q' and r'. Recall that the dual space of $L^p(K)$ is $L^{p'}(K)$ for $1 \le p < \infty$. Also recall that the adjoint $T^* : Y' \to X'$ of a bounded linear map T between two Banach spaces X and Y is defined by (see Folland [61] for more details)

$$[T^*\ell](x) = \ell(T(x)), \quad \ell \in Y', \quad x \in X.$$

The next lemma provides a standard method to establish the boundedness of linear operators between L^p spaces; see Stein [133, p:280]

Lemma 1.4 *(TT^* method) Let T be a linear operator defined on a dense subset of L^2, and with a formal adjoint T^*. Then the following are equivalent:*
(i) $\|T\|_{L^2\to L^p} \leq A$,
(ii) $\|T^*\|_{L^{p'}\to L^2} \leq A$,
(iii) $\|TT^*\|_{L^{p'}\to L^p} \leq A^2$.

To study low regularity solutions of PDEs, we need to define the solution in a distributional sense. We introduce the following test function spaces

$$C^\infty(K) = \{f : K \to \mathbb{C} : f \text{ is infinitely differentiable}\}, \quad K = \mathbb{R} \text{ or } \mathbb{T},$$

$$C_0^\infty(\mathbb{R}) = \{f \in C^\infty(\mathbb{R}) : f \text{ is compactly supported}\},$$

$$\mathcal{S}(\mathbb{R}) = \left\{f \in C^\infty(\mathbb{R}) : P_{m,n}(f) := \left\|\langle x\rangle^m f^{(n)}(x)\right\|_{L^\infty} < \infty, m, n \geq 0\right\}.$$

Recall that these spaces are locally convex topological vector spaces. The topology on C^∞ is given by uniform convergence of each derivative on compact sets. Similarly, we say f_j converges to f in C_0^∞ if there is a compact set F containing the support of each f_j and f, and that f_j and all of its derivatives converges to f and its derivatives uniformly on F. We say f_j converges to f in $\mathcal{S}(\mathbb{R})$ if $P_{m,n}(f - f_j)$ converges to 0 for each $m, n \geq 0$.

Note that the dual space S' of a topological vector space S is defined analogously as the space of linear continuous maps from S to $\mathbb{C}$. We define the space of distributions on $\mathbb{R}$, $\mathcal{D}(\mathbb{R})$, as the dual of $C_0^\infty(\mathbb{R})$, and the space of periodic distributions, $\mathcal{D}(\mathbb{T})$, as the dual of $C^\infty(\mathbb{T})$. We also define the space of tempered distributions, $\mathcal{S}'(\mathbb{R})$, as the dual of $\mathcal{S}(\mathbb{R})$. We refer the reader to Folland [61] for the basic properties of distributions. We denote the action of a distribution u on a test function ϕ by $u(\phi)$. If u is an L^p function, then we have (see Exercise 1.2)

$$u(\phi) = \int_{\mathbb{R}} u(x)\phi(x)\, dx.$$

We define the Fourier transform for functions in $L^1(\mathbb{R})$ as

$$\mathcal{F}f(\xi) = \widehat{f}(\xi) = \frac{1}{\sqrt{2\pi}} \int_{\mathbb{R}} f(x)e^{-i\xi x}dx, \quad \xi \in \mathbb{R}.$$

Recall that $\widehat{f}$ is a continuous bounded function on $\mathbb{R}$ that decays to zero at infinity. In the case when $\widehat{f} \in L^1(\mathbb{R})$, one has the inversion formula

$$f(x) = \mathcal{F}^{-1}\widehat{f}(x) = \frac{1}{\sqrt{2\pi}} \int_{\mathbb{R}} e^{ix\xi} \widehat{f}(\xi) d\xi, \quad x \in \mathbb{R}.$$

Similarly, for $f \in L^1(\mathbb{T})$, we define the Fourier series as

$$\mathcal{F} f(k) = \widehat{f}(k) = \frac{1}{2\pi} \int_0^{2\pi} f(x) e^{-ikx} dx, \quad k \in \mathbb{Z}.$$

In the case $\widehat{f} \in \ell^1$, we have

$$f(x) = \mathcal{F}^{-1}\widehat{f}(x) = \sum_{k \in \mathbb{Z}} e^{ikx} \widehat{f}(k).$$

We have the Poisson summation formula ; see e.g. Folland [61]

$$\sum_{k=-\infty}^{\infty} f(x + 2\pi k) = \frac{1}{\sqrt{2\pi}} \sum_{k=-\infty}^{\infty} \widehat{f}(k) e^{ixk}, \quad f \in \mathcal{S}(\mathbb{R}). \tag{1.2}$$

For $f, g \in L^1(\mathbb{R})$, we have the Fourier multiplication formula

$$\int_{\mathbb{R}} f(x) \widehat{g}(x)\, dx = \int_{\mathbb{R}} \widehat{f}(x) g(x)\, dx. \tag{1.3}$$

This leads to Parseval's identity

$$\langle f, g \rangle = \int_{\mathbb{R}} \overline{f(x)} g(x)\, dx = \int_{\mathbb{R}} \overline{\widehat{f}(\xi)} \widehat{g}(\xi)\, d\xi = \left\langle \widehat{f}, \widehat{g} \right\rangle, \quad f, g \in L^2(\mathbb{R}),$$

and, in particular, we have Plancherel's theorem

$$\|f\|_{L^2(\mathbb{R})} = \left\|\widehat{f}\right\|_{L^2(\mathbb{R})}.$$

Similar formulas hold in the case of Fourier series. Interpolating Plancherel's theorem with the inequality

$$\left\|\widehat{f}\right\|_{L^\infty} \le \|f\|_{L^1}$$

yields the Hausdorff–Young theorem

$$\left\|\widehat{f}\right\|_{L^{p'}} \le \|f\|_{L^p}, \quad 1 \le p \le 2, \quad \frac{1}{p} + \frac{1}{p'} = 1.$$

By noting that $\mathcal{F}$ is a continuous bijection from $\mathcal{S}(\mathbb{R})$ to $\mathcal{S}(\mathbb{R})$, one can extend the definition of the Fourier transform to the space $\mathcal{S}'(\mathbb{R})$ by the formula

$$\mathcal{F} u(\phi) := u(\mathcal{F} \phi).$$

Similarly, one can define the Fourier series of periodic distributions as

$$\mathcal{F}u(k) = \frac{1}{2\pi} u(e^{-ik\cdot}).$$

Most of the basic properties of $\mathcal{F}$ can be extended to the distributional definitions, see [61]. In particular, both distributional definitions agree with the usual definition for $u \in L^1(K)$.

For $s > 0$, we often use the operators $D^s = (-\Delta)^{\frac{s}{2}}$ given on the Fourier side as

$$\widehat{D^s f}(\xi) = |\xi|^s \widehat{f}(\xi).$$

Similarly, we have the operators J^s given on the Fourier side as

$$\widehat{J^s f}(\xi) = \langle \xi \rangle^s \widehat{f}(\xi).$$

We define the L^2 based Sobolev spaces, for $s \in \mathbb{R}$

$$H^s(\mathbb{R}) = \{ f \in \mathcal{S}'(\mathbb{R}) : \|f\|_{H^s(\mathbb{R})} := \|J^s f\|_{L^2} < \infty \},$$

where $\langle \xi \rangle := \sqrt{1 + |\xi|^2}$. Similarly, for $s \in \mathbb{R}$, we have

$$H^s(\mathbb{T}) = \{ f \in \mathcal{D}(\mathbb{T}) : \|f\|_{H^s(\mathbb{T})} := \|J^s f\|_{L^2} < \infty \}.$$

The homogenous Sobolev spaces $\dot{H}^s$ are defined analogously with D^s instead of J^s.

Recall that C_0^∞ functions are dense in $H^s(K)$ for any s and in $L^p(K)$ for $1 \le p < \infty$. Note that for $\alpha > 0$, $D^\alpha : H^s \to H^{s-\alpha}$, and for any α, $J^\alpha : H^s \to H^{s-\alpha}$. Also note that

$$\partial_x f = HDf = DHf,$$

where H is the Hilbert transform

$$\widehat{Hf}(\xi) = i\text{sign}(\xi)\widehat{f}(\xi),$$

which is bounded in $H^s(\mathbb{R})$ for any s, and in L^p for $1 < p < \infty$.

We collect some basic properties of Sobolev spaces in the following lemmas (see Exercises 1.3 and 1.4):

Lemma 1.5 *(Sobolev embedding) For $K = \mathbb{R}$ or $\mathbb{T}$, we have*

$$\|f\|_{L^p(K)} \lesssim \|f\|_{\dot{H}^s(K)}, \quad s = \frac{1}{2} - \frac{1}{p}, \quad 2 < p < \infty,$$

$$\|f\|_{L^\infty(K)} \lesssim \|f\|_{H^s(K)}, \quad s > \frac{1}{2}.$$

Lemma 1.6 *(Algebra property) For $K = \mathbb{R}$ or $\mathbb{T}$, and $s > \frac{1}{2}$, we have*

$$\|fg\|_{H^s(K)} \lesssim \|f\|_{H^s(K)}\|g\|_{H^s(K)}.$$

We also need basic definitions and theorems from the Littlewood–Paley theory. We start with Littlewood–Paley projections

$$\widehat{P_k f}(\xi) = \varphi(2^{-k}\xi)\widehat{f}(\xi),$$

where φ is a smooth cut-off function supported in $\{\xi : \frac{1}{2} \leq |\xi| \leq 2\}$ with the property

$$\sum_k \varphi(2^{-k}\xi) = 1, \quad \text{for each } \xi \neq 0.$$

We use the same definition for functions on the torus. Similarly, $P_{\geq k}$ is given as

$$\widehat{P_{\geq k} f}(\xi) = \widehat{f}(\xi)\sum_{j=k}^{\infty} \varphi(2^{-j}\xi),$$

and $P_{<k} = Id - P_{\geq k}$. It is important to note the uniform bounds

$$|P_k f(x)|, |P_{<k} f(x)| \leq C\, Mf(x). \tag{1.4}$$

Here M is the Hardy–Littlewood maximal function

$$Mf(x) = \sup_{r>0} \frac{1}{|B(x,r)|}\int_{B(x,r)} |f(y)|\, dy,$$

where $B(x,r)$ is the ball of radius r centered at x, and $|B(x,r)|$ is the volume of $B(x,r)$. Both inequalities follow by majorizing the inverse Fourier transform of the cutoff function by a sum of characteristic functions of balls. The Hardy–Littlewood theorem states that for any $1 < p \leq \infty$

$$\|Mf\|_{L^p} \leq C_p \|f\|_{L^p},$$

both on $\mathbb{R}$ and $\mathbb{T}$. This implies uniform L^p boundedness of the Littlewood–Paley projections via (1.4).

We also have the following characterization of the Sobolev spaces in terms of Littlewood–Paley projections (see Exercise 1.5)

$$\|f\|_{H^s}^2 \approx \|P_{<1} f\|_{L^2}^2 + \sum_{k=1}^{\infty} 2^{2ks}\|P_k f\|_{L^2}^2. \tag{1.5}$$

Also recall that for $f \in \mathcal{S}(\mathbb{R})$, we have

$$P_{<1} f + \sum_{k=1}^{n} P_k f \to f \text{ in } \mathcal{S}(\mathbb{R}), \tag{1.6}$$

as $n \to \infty$.

We finish this section by stating some additional results on Sobolev spaces. The proofs of these results are more involved, and Taylor [146] is a good reference for these and related inequalities.

We first state the Gagliardo–Nirenberg inequality [117]:

Theorem 1.7 *Let $f : \mathbb{R} \to \mathbb{C}$. Fix $1 \le p, q, r \le \infty$ and a natural number m. Suppose also that a real number α and a natural number j are such that*

$$\frac{1}{p} = j + \left(\frac{1}{r} - m\right)\alpha + \frac{1-\alpha}{q}$$

and

$$\frac{j}{m} \le \alpha \le 1.$$

Then

$$\|D^j f\|_{L^p} \lesssim \|D^m f\|_{L^r}^{\alpha} \|f\|_{L^q}^{1-\alpha}.$$

We will mainly use the following corollary:

Corollary 1.8 *For any $j, k \in \mathbb{N}$, we have*

$$\left\|\partial_x^j f\, \partial_x^k f\right\|_{L^2} \lesssim \|f\|_{H^{k+j}} \|f\|_{L^\infty}.$$

Proof We can assume that $j, k \ge 1$. Let $m = j + k$. Let

$$p_1 = \frac{2m}{j}, \quad p_2 = \frac{2m}{k}, \quad \alpha_1 = \frac{j}{m}, \quad \alpha_2 = \frac{k}{m}.$$

Note that in this case $\frac{1}{p_1} + \frac{1}{p_2} = \frac{1}{2}$ and $\alpha_1 + \alpha_2 = 1$. By Hölder's and Gagliardo–Nirenberg's inequalities, and the L^p boundedness of the Hilbert transform, we have

$$\begin{aligned}
\left\|\partial_x^j f\, \partial_x^k f\right\|_{L^2} &\le \left\|\partial_x^j f\right\|_{L^{p_1}} \left\|\partial_x^k f\right\|_{L^{p_2}} \\
&\lesssim \left\|D^j f\right\|_{L^{p_1}} \left\|D^k f\right\|_{L^{p_2}} \lesssim \|D^m f\|_{L^2}^{\alpha_1+\alpha_2} \|f\|_{L^\infty}^{2-(\alpha_1+\alpha_2)} \\
&\le \|f\|_{H^m} \|f\|_{L^\infty}.
\end{aligned}$$

□

The following is a standard commutator estimate that we use in Chapter 3 to establish the wellposedness theory for some dispersive PDEs. For the proof, see Kato–Ponce [83], Kenig–Ponce–Vega [88], and Taylor [146, page 106].

Lemma 1.9 *For $s \in (0, 1)$, we have*

$$\|J^s(fg) - fJ^s g\|_{L^2} \lesssim \|f\|_{H^s} \|g\|_{L^\infty}.$$

For $s > 1$, we have

$$\left\|J^s(fg) - fJ^s g\right\|_{L^2} \lesssim \|f\|_{H^s}\|g\|_{L^\infty} + \|f_x\|_{L^\infty}\|g\|_{H^{s-1}}.$$

We state the following version of fractional Leibniz rule; see Taylor [146].

Lemma 1.10 *For $s \geq 0$*

$$\frac{1}{2} = \frac{1}{p_1} + \frac{1}{q_1} = \frac{1}{p_2} + \frac{1}{q_2}, \quad 2 < p_1, q_2 \leq \infty,$$

we have

$$\|fg\|_{H^s} \lesssim \|f\|_{L^{p_1}}\|J^s g\|_{L^{q_1}} + \|J^s f\|_{L^{p_2}}\|g\|_{L^{q_2}}.$$

Finally, we have the following L^r version of this lemma; see Taylor [146].

Lemma 1.11 *For $s \geq 0$, $1 < r < \infty$, we have*

$$\|J^s(fg)\|_{L^r} \lesssim \|f\|_{L^{p_1}}\|J^s g\|_{L^{q_1}} + \|J^s f\|_{L^{p_2}}\|g\|_{L^{q_2}},$$

where $p_j, q_j \in (1, \infty)$, and

$$\frac{1}{r} = \frac{1}{p_1} + \frac{1}{q_1} = \frac{1}{p_2} + \frac{1}{q_2}.$$

Exercises

1.1 Prove that

$$\|f * g\|_{L^p} \leq \|f\|_{L^p}\|g\|_{L^1}, \quad \text{and}$$

$$\|f * g\|_{L^\infty} \leq \|f\|_{L^p}\|g\|_{L^{p'}}.$$

Derive Young's inequality (1.1) using these bounds and Theorem 1.1.

1.2 Fix $p \in [1, \infty]$. Prove that any $f \in L^p(\mathbb{R})$ is a tempered distribution with the action given by

$$f(\phi) = \int_{\mathbb{R}} f(x)\phi(x)\,dx, \quad \phi \in \mathcal{S}(\mathbb{R}).$$

1.3 (a) Consider the tempered distribution $|x|^{-\alpha}$, $0 < \alpha < 1$ on $\mathbb{R}$. Prove that

$$\mathcal{F}\left(|\cdot|^{-\alpha}\right)(\xi) = c_\alpha|\xi|^{\alpha-1},$$

where c_α is a constant depending only on α.

(b) Use the Hardy–Littlewood–Sobolev theorem, part (a), and duality to prove the Sobolev embedding theorem on $\mathbb{R}$

$$\|f\|_{L^p(\mathbb{R})} \leq C\|D^s f\|_{L^2(\mathbb{R})}$$

with $\frac{1}{2} = \frac{1}{p} + s$ and $2 < p < \infty$.

(c) Use part (b) to prove the following version of the Gagliardo–Nirenberg inequality on $\mathbb{R}$, for $p \in [2, \infty]$ and $\theta = \frac{1}{2} - \frac{1}{p}$

$$\|f\|_{L^p} \leq C\|f_x\|_{L^2}^{\theta}\|f\|_{L^2}^{1-\theta}. \tag{1.7}$$

(d) Give an alternative proof of (1.7) by interpolating the bounds for $p = 2$ and $p = \infty$. For the latter bound, express f^2 using the fundamental theorem of calculus.

1.4 Prove the algebra of Sobolev spaces stated in Lemma 1.6.

1.5 Prove (1.5) using Plancherel's theorem.

1.6 (Gronwall's inequality) Assume that for a.e. $t \in [0, T]$, we have

$$f(t) \leq A + \int_0^t g(\tau)f(\tau)d\tau$$

for some $A \geq 0$ and some nonnegative functions f and g such that $fg \in L^1([0, T])$. Prove that

$$f(t) \leq A \exp\left(\int_0^t g(\tau)d\tau\right), \quad t \in [0, T].$$

1.7 Consider the linear Schrödinger equation on $\mathbb{R}$

$$\begin{cases} iu_t + u_{xx} = 0, x \in \mathbb{R}, t \in \mathbb{R}, \\ u(0, \cdot) = g(\cdot) \in H^s(\mathbb{R}). \end{cases} \tag{1.8}$$

(a) For each $t \in \mathbb{R}$, define $u(t, \cdot)$ as a tempered distribution by the formula

$$u(t, \cdot) = \mathcal{F}^{-1}\left[e^{-it\xi^2}\mathcal{F}g(\xi)\right](\cdot).$$

Prove that $u \in C_t^0 H_x^s(\mathbb{R} \times \mathbb{R})$ and that $\|u(t, \cdot)\|_{H^s(\mathbb{R})}$ is constant.

(b) Prove that if g_n converges to g in H^s, then, for each t, $u_n(t, \cdot)$ converges to $u(t, \cdot)$ in H^s.

(c) Prove that u solves (1.8) in the sense of distributions, i.e.

$$u(-i\phi_t + \phi_{xx}) = 0,$$

for all $\phi \in \mathcal{S}(\mathbb{R}^2)$.

(d) Prove that if $v \in C_t^0 H_x^s(\mathbb{R} \times \mathbb{R})$, $v(0, x) = g(x)$, and v solves (1.8) in the sense of distributions, then $v = u$.

1.8 (a) Let $\phi \in \mathcal{S}(\mathbb{R})$ and $z \in \mathbb{C}\backslash\{0\}$ with a nonnegative real part. Then

$$\int_{\mathbb{R}} e^{-z|x|^2}\widehat{\phi}(x)dx = \frac{1}{\sqrt{2z}}\int_{\mathbb{R}} e^{-\frac{|x|^2}{4z}}\phi(x)dx.$$

(b) Assuming that $g \in \mathcal{S}(\mathbb{R})$, express the solution of (1.8) as a convolution of the tempered distribution $\frac{1}{\sqrt{4\pi i t}} e^{i\frac{|x|^2}{4t}}$ with g.

(c) Similarly, prove that the solution of the linear Schrödinger equation on $\mathbb{R}^n$

$$\begin{cases} iu_t + \Delta u = 0, x \in \mathbb{R}^n, t \in \mathbb{R}, \\ u(0,\cdot) = g(\cdot) \in \mathcal{S}(\mathbb{R}^n) \end{cases} \tag{1.9}$$

is given by

$$e^{it\Delta}g = \frac{1}{(4\pi i t)^{\frac{n}{2}}} \int_{\mathbb{R}^n} e^{i\frac{|x-y|^2}{4t}} g(y)\,dy.$$

(d) Conclude that the following dispersive estimate holds

$$\left\|e^{it\Delta}g\right\|_{L_x^\infty(\mathbb{R}^n)} \le \frac{1}{(4\pi|t|)^{\frac{n}{2}}} \|g\|_{L^1(\mathbb{R}^n)}.$$

1.9 Let $f \in C_0^\infty(\mathbb{R})$, $\phi \in C^\infty$, and $\phi'(x) \neq 0$ for any x in the support of f. Then

$$I(\lambda) = \int_{\mathbb{R}} e^{i\lambda\phi(x)} f(x)dx = O(\lambda^{-k}), \text{ as } \lambda \to \infty$$

for any $k \in \mathbb{Z}^+$.

1.10 (Van der Corput lemma) Let $\phi : \mathbb{R} \to \mathbb{R}$ be C^∞.

(a) Assume that for some $k \in \mathbb{Z}^+$, we have $|\phi^{(k)}(x)| \ge 1$ for any $x \in [a,b]$, with $\phi'(x)$ monotonic when $k = 1$. Then

$$\left|\int_a^b e^{i\lambda\phi(x)}dx\right| \le c_k \lambda^{-\frac{1}{k}},$$

where the constant c_k is independent of a and b.

Hint: The case $k = 1$ follows from an integration by parts noting that ϕ'' does not change sign. For the general case, use induction, noting that $\phi^{(k-1)}$ can vanish at most at a single point, and decompose $[a,b]$ into two disjoint sets appropriately.

(b) Under the hypothesis of part (a), prove that

$$\left|\int_a^b e^{i\lambda\phi(x)} f(x)dx\right| \le c_k \lambda^{-\frac{1}{k}} \left(\|f\|_{L^\infty} + \|f'\|_{L^1}\right).$$

1.11 (Bernstein's inequality) Prove that if P is a trigonometric polynomial of degree N, then

$$\left\|P'\right\|_{L^p} \lesssim N\|P\|_{L^p}, \quad 1 \le p \le \infty.$$

Hint: Express P' as

$$P' = P * K_N,$$

where K_N is a suitable convolution kernel with the property $\|K_N\|_{L^1} \lesssim N$, see Katznelson [84].

1.12 Using the Poisson summation formula, (1.2) prove that the convolution kernel φ_k of P_k on the torus satisfies the bound

$$|\varphi_k(x)| \leq C \frac{2^k}{(1+2^k|x|)^2}, \quad k \in \mathbb{N}, |x| \leq \frac{1}{2}.$$

2
Linear dispersive equations

In this chapter, we study the basic properties of linear dispersive PDEs. A linear PDE is called dispersive if frequency localized data evolves without changing its shape with a velocity depending on the frequency. For example, note that the plane wave

$$u(t,x) = e^{i(\xi x - \omega t)}$$

is a solution of the linear Schrödinger equation

$$iu_t + u_{xx} = 0$$

if and only if the so-called dispersion relation, $\omega = \xi^2$, holds. Note that ω is a real-valued function of the frequency.[1] Denoting the phase velocity by $v = \frac{\omega}{\xi}$, one can write the solution as

$$u(t,x) = e^{i\xi(x - v(\xi)t)} = u(0, x - v(\xi)t),$$

and conclude that the wave travels with velocity $v(\xi)$. In particular, large frequency data travel faster than smaller ones. A related notion is the group velocity, $\omega'(\xi)$, which describes how a frequency localized bump function around ξ moves, see e.g. Griffiths [71]. To see why the group velocity is different than the phase velocity, let $g(x) = u(0,x)$ be concentrated around a frequency ξ_0. We can write the solution as

$$u(t,x) = \frac{1}{\sqrt{2\pi}} \int_{-\infty}^{\infty} \widehat{g}(\xi) e^{-it\xi^2} e^{ix\xi} d\xi.$$

[1] A commonly used defining criteria for dispersive equations is that $\omega(\xi)$ is a real-valued function of ξ and $\frac{d^2\omega}{d\xi^2} \not\equiv 0$. Under this criterion, the transport equation $u_t = u_x$ and the wave equation $u_{tt} = u_{xx}$ are not dispersive.

Approximating $\omega(\xi) = \xi^2$ around ξ_0

$$\begin{aligned}\xi^2 &= \omega(\xi_0) + (\xi - \xi_0)\omega'(\xi_0) + O\left((\xi - \xi_0)^2\right)\\ &= 2\xi_0\xi - \xi_0^2 + O\left((\xi - \xi_0)^2\right),\end{aligned}$$

and ignoring the error term, we have

$$\begin{aligned}u(t,x) &\approx e^{it\xi_0^2}\frac{1}{\sqrt{2\pi}}\int_{-\infty}^{\infty}\widehat{g}(\xi)e^{-2it\xi_0\xi}e^{ix\xi}d\xi\\ &= e^{it\xi_0^2}g(x - 2\xi_0 t).\end{aligned}$$

This suggests that the bump function moves with the group velocity $\omega'(\xi_0) = 2\xi_0$, which is twice the phase velocity at ξ_0. In general, the convexity of ω determines the relation between the phase and the group velocities. For example, for $\omega(\xi) = \xi^\alpha$, the group velocity is α times the phase velocity.

Comparing the plain wave solutions above with the plain wave solutions of the heat equation $u_t - u_{xx} = 0$, we see that $\omega = -i\xi^2$ is complex valued. Therefore, each nonzero mode decays exponentially in time. Spatially localized solutions of dispersive equations also decay in time (at a slower rate) due to a more involved reason. When no boundary conditions are imposed, different frequency components of the data evolve with different velocities, and hence spread out in space as time increases. This causes the solution to decay in time at a polynomial rate. This can also be understood by noting the L^2 conservation $\|u(t)\|_{L^2} = \|u(0)\|_{L^2}$ for the solutions of the linear Schrödinger equation. A spatially localized smooth bump of height $\sqrt{k}$ and base $1/k$ would spread out to an interval of length kt at time t and hence by the L^2 conservation one expects the solution to decay to the height $\frac{1}{\sqrt{kt}}$.

In contrast, on bounded domains such as the torus $\mathbb{T}$ the solution cannot spread out, but instead different frequency components rotate around the torus with different velocities. This makes time averages smoother because of a subtle cancellation between different frequency components.

Another characterization of dispersive equations comes from the curvature of the support of the space-time Fourier transform (we usually denote by (τ, ξ) the dual variables of (t, x)) of their solutions. For example, taking the Fourier transform of the solution of the equation $iu_t + u_{xx} = 0$, we see that $u(\tau, \xi)$ is supported on the hypersurface $\tau = \xi^2$, which has a nonvanishing Gaussian curvature. In this light, the linear wave equation in higher dimensions is also dispersive since the Fourier support of the solution is the cone $\tau = |\xi|$, which has $d - 1$ nonvanishing principle curvatures.

The methods we present in this book to establish wellposedness of a nonlinear dispersive PDE treat the equation as a nonlinear perturbation of the corresponding linear equation. As such, these methods rely heavily on the dispersive estimates for the linear group of the PDEs, which we develop in this chapter. We consider linear dispersive equations of the form

$$\begin{cases} iu_t = Lu, & t \in \mathbb{R}, \ x \in \mathbb{T} \text{ or } \mathbb{R}, \\ u(0,\cdot) = g(\cdot) \in H^s, \end{cases}$$

where L is a linear symmetric differential operator with constant coefficients. We concentrate on the model examples:

$$\begin{aligned} &L = -\partial_{xx}, \quad \text{linear Schrödinger equation,} \\ &L = -i\partial_{xxx}, \quad \text{Airy equation (linear part of the KdV equation),} \\ &L = D^{2\alpha}, \quad \alpha \in (1/2, 1), \quad \text{linear fractional Schrödinger equation.} \end{aligned}$$

The estimates we obtain can be extended to higher order dispersive PDEs, and in some cases to higher dimensions. We first briefly discuss the case of the linear Schrödinger evolution on the real line to motivate discussion of the periodic case and of the Airy and fractional Schrödinger equations. Estimates on the real line rely on standard oscillatory integral techniques such as the nonstationary phase (Exercise 1.9) and the Van der Corput lemma (Exercise 1.10) along with the Hardy–Littlewood–Sobolev inequality and TT^* method.

2.1 Estimates on the real line

Consider the linear Schrödinger equation on $\mathbb{R}$:

$$\begin{cases} iu_t + u_{xx} = 0, & t \in \mathbb{R}, \ x \in \mathbb{R}, \\ u(0,\cdot) = g(\cdot) \in H^s(\mathbb{R}). \end{cases} \tag{2.1}$$

The solution of the initial value problem $u(t,x) = e^{it\partial_{xx}}g$ can be expressed by employing the Fourier transform (see Exercises 1.7 and 1.8) as follows

$$\begin{aligned} e^{it\partial_{xx}}g &= \mathcal{F}^{-1}\left(e^{-it\xi^2}\widehat{g}(\xi)\right)(x) \\ &= \frac{1}{\sqrt{4\pi it}} \int_{\mathbb{R}} e^{i\frac{|x-y|^2}{4t}} g(y)dy. \end{aligned}$$

This immediately implies the conservation of H^s norms

$$\left\|e^{it\partial_{xx}}g\right\|_{H^s_x} = \|g\|_{H^s}, \quad s \in \mathbb{R}.$$

We also note that, by the dominated convergence theorem, the solution belongs to $C^0_t H^s_x$ for H^s data. The formula above also implies the $L^1 \to L^\infty$ dispersive estimate (also see Exercise 1.8)

$$\left\| e^{it\partial_{xx}} g \right\|_{L^\infty_x} \le \frac{1}{\sqrt{4\pi|t|}} \|g\|_{L^1}.$$

Interpolating the $L^1 \to L^\infty$ and $L^2 \to L^2$ estimates above via the Riesz–Thorin interpolation theorem, one obtains for $r \ge 2$

$$\left\| e^{it\partial_{xx}} g \right\|_{L^r_x} \le \frac{1}{(4\pi|t|)^{\frac{1}{2}-\frac{1}{r}}} \|g\|_{L^{r'}}. \tag{2.2}$$

Using (2.2), the TT^* method, and the Hardy–Littlewood–Sobolev inequality, one obtains the following space-time bound known as a Strichartz estimate

$$\left\| e^{it\partial_{xx}} g \right\|_{L^q_t L^r_x} \lesssim \|g\|_{L^2}, \quad \frac{2}{q} + \frac{1}{r} = \frac{1}{2}, \quad r \ge 2. \tag{2.3}$$

This estimate is optimal as it can be seen by scaling; if u solves (2.1), then so does $u(\lambda^2 t, \lambda x)$ with data $g(\lambda x)$ for any $\lambda \in \mathbb{R}$.

Strichartz estimates were first obtained by Strichartz in [136] utilizing the Fourier restriction theorem of Stein–Tomas [148]. The proof outlined above was discovered later by Ginibre–Velo [67]. Strichartz estimates hold in every dimension with the scaling relation in $\mathbb{R}^n$, $n \ge 2$, being

$$\frac{2}{q} + \frac{n}{r} = \frac{n}{2}, \quad 2 \le r \le \frac{2n}{n-2}, \quad (n, r) \ne (2, \infty).$$

The proof is identical to the proof sketched above with the exception of the endpoint $r = \frac{2n}{n-2}$, $n \ge 3$; see Exercise 2.1. This endpoint result was obtained later by Keel–Tao in [85]. It is also known that the endpoint result $(n, r) = (2, \infty)$ cannot hold, see Montgomery-Smith [114]. However, it holds for spherically symmetric initial data, see Tao [141] and Stefanov [131].

One can also obtain space-time estimates for the inhomogenous problem; see Ginibre–Velo [67], Cazenave [30], and Linares–Ponce [105]; also Exercise 2.1.

For the wave equation

$$\begin{cases} u_{tt} - \Delta u = 0, & x \in \mathbb{R}^n, \ t \in \mathbb{R} \\ u(0, x) = f(x), & u_t(0, x) = g(x), \end{cases} \tag{2.4}$$

where the dispersion is weaker (see Exercise 2.2), the Strichartz estimates hold with a derivative loss

$$\|u\|_{L^q_t L^r_x} \lesssim \|f\|_{\dot{H}^s(\mathbb{R}^n)} + \|g\|_{\dot{H}^{s-1}(\mathbb{R}^n)},$$

for any $n \ge 2$ provided that

$$\frac{2}{q} + \frac{n-1}{r} \le \frac{n-1}{2}, \quad 2 \le r \le \frac{2n-2}{n-3}, \quad (n, r) \ne (3, \infty),$$

and

$$\frac{1}{q} + \frac{n}{r} = \frac{n}{2} - s,$$

see Strichartz [136] and Keel–Tao [85].

Strichartz estimates are a very useful tool in the wellposedness theory of nonlinear Schrödinger equations with power type nonlinearities. We will present them in more detail for the Airy evolution on $\mathbb{R}$

$$\begin{cases} u_t + u_{xxx} = 0, & t \in \mathbb{R}, \ x \in \mathbb{R}, \\ u(0, \cdot) = g(\cdot) \in H^s(\mathbb{R}). \end{cases} \tag{2.5}$$

For $g \in \mathcal{S}(\mathbb{R})$, the solution $e^{-t\partial_{xxx}} g$ can be expressed as a convolution with a distribution by employing the Fourier transform as in Exercises 1.7 and 1.8

$$e^{-t\partial_{xxx}} g = \mathcal{F}^{-1}\left(e^{it\xi^3} \widehat{g}(\xi)\right)(x) = A_t * g(x),$$

where A_t is the Airy kernel

$$A_t(x) = \frac{1}{\sqrt{2\pi}} \mathcal{F}^{-1} e^{it\xi^3}.$$

Moreover, we will see that for $t \neq 0$, A_t is a bounded function of x. From now on, we use the notation W_t to denote the propagator, $e^{-t\partial_{xxx}}$, of the Airy equation. The Fourier representation immediately implies the conservation of H^s norms

$$\|W_t g\|_{H^s(\mathbb{R})} = \|g\|_{H^s(\mathbb{R})}, \quad s \in \mathbb{R}.$$

One can also obtain $L^1 \to L^\infty$ and Strichartz estimates for the evolution $W_t g$; see Kenig–Ponce–Vega [88], Proposition 2.1, and Theorem 2.4 below. The proofs are more involved since there is no explicit formula for the Airy kernel. It is important to note that the dispersion is stronger for large frequencies and weaker for small frequencies compared to the linear Schrödinger evolution. In $L^1 \to L^\infty$ estimates, this allows one to have some derivative gain. It also causes a weaker bound as $t \to \infty$ and a stronger bound as $t \to 0$ as the following proposition establishes.

Proposition 2.1 *For any $\alpha \in [0, \frac{1}{2}]$, and $t \in \mathbb{R}$, we have*

$$\|D^\alpha W_t g\|_{L^\infty_x} \leq C_\alpha |t|^{-(\alpha+1)/3} \|g\|_{L^1}.$$

We start the proof with the following oscillatory integral estimate:

Lemma 2.2 *Let ϕ be a smooth cutoff function supported in $\{\xi \in \mathbb{R} : \frac{1}{2} \leq |\xi| \leq 2\}$. Let*

$$\Phi(t, x) := \int_{\mathbb{R}} e^{it\xi^3 + ix\xi} \phi(\xi) d\xi.$$

Then for each $x, t \in \mathbb{R}$, we have

$$|\Phi(t,x)| \lesssim \langle t \rangle^{-\frac{1}{2}}.$$

Moreover, if $|x| \not\approx |t|$, then

$$|\Phi(t,x)| \lesssim \min\left(|x|^{-2}, |t|^{-2}\right).$$

Proof The first bound follows from the trivial inequality $|\Phi(t,x)| \lesssim 1$ and the Van der Corput Lemma (Exercise 1.10) by noting the following estimate for the second derivative of the phase function on the support of ϕ

$$\left|\partial_\xi^2\left(\xi^3 t + \xi x\right)\right| = 6|\xi t| \gtrsim |t|.$$

The second bound follows from two integration by parts using the bound

$$\left|\partial_\xi\left(\xi^3 t + \xi x\right)\right| = \left|3\xi^2 t + x\right| \gtrsim \max(|x|, |t|),$$

which is valid in the region $|x| \not\approx |t|$, $\frac{1}{2} \le |\xi| \le 2$.

Indeed

$$\begin{aligned}
\Phi(t,x) = \int_{\mathbb{R}} \frac{\phi(\xi)}{3i\xi^2 t + ix} \partial_\xi e^{it\xi^3 + ix\xi} d\xi \\
&= -\int_{\mathbb{R}} \partial_\xi\left(\frac{\phi(\xi)}{3i\xi^2 t + ix}\right) e^{it\xi^3 + ix\xi} d\xi \\
&= -\int_{\mathbb{R}} \partial_\xi\left(\frac{1}{3\xi^2 t + x}\partial_\xi\left(\frac{\phi(\xi)}{3\xi^2 t + x}\right)\right) e^{it\xi^3 + ix\xi} d\xi.
\end{aligned}$$

Therefore

$$\begin{aligned}
|\Phi(t,x)| &\lesssim \sup_{|\xi|\approx 1}\left|\partial_\xi\left(\frac{1}{3\xi^2 t + x}\partial_\xi\left(\frac{\phi(\xi)}{3\xi^2 t + x}\right)\right)\right| \\
&\lesssim \sup_{|\xi|\approx 1}\left(\frac{1}{(3\xi^2 t + x)^2} + \frac{|t|}{|3\xi^2 t + x|^3} + \frac{t^2}{(3\xi^2 t + x)^4}\right) \\
&\lesssim \frac{1}{\max(|x|^2, |t|^2)}.
\end{aligned}$$

□

Proof of Proposition 2.1 It suffices to prove the proposition for $g \in \mathcal{S}(\mathbb{R})$. Fix $\alpha \in [0, \frac{1}{2}]$ and note that

$$D^\alpha W_t g = \frac{1}{\sqrt{2\pi}} D^\alpha (A_t * g) = \frac{1}{\sqrt{2\pi}} (D^\alpha A_t) * g.$$

For $t > 0$, we have the scaling relation

$$|D^\alpha A_t(x)| = t^{-(\alpha+1)/3}\left|D^\alpha A_1(x/t^{1/3})\right|.$$

Therefore, by Young's inequality, it suffices to prove that $|D^\alpha A_1(x)|$ is a bounded function. Given $f \in \mathcal{S}(\mathbb{R})$, we have (by (1.6))

$$\begin{aligned} D^\alpha A_1(f) &= \mathcal{F}^{-1}\left(|\xi|^\alpha e^{i\xi^3}\right)(f) \\ &= \mathcal{F}^{-1}\left(|\xi|^\alpha e^{i\xi^3}\right)(P_{<1}f) + \lim_{n\to\infty}\sum_{k=1}^{n} \mathcal{F}^{-1}\left(|\xi|^\alpha e^{it\xi^3}\right)(P_k f) \\ &= \mathcal{F}^{-1}\left(|\xi|^\alpha e^{i\xi^3}\psi(\xi)\right)(f) + \lim_{n\to\infty}\sum_{k=1}^{n} \mathcal{F}^{-1}\left(|\xi|^\alpha e^{i\xi^3}\varphi(\xi 2^{-k})\right)(f). \end{aligned}$$

Here ψ is a smooth, compactly supported bump function, and φ is the function in the definition of the Littlewood–Paley projections. This identifies $D^\alpha A_1$ as the distributional limit:

$$D^\alpha A_1 = \mathcal{F}^{-1}\left(|\xi|^\alpha e^{i\xi^3}\psi(\xi)\right) + \lim_{n\to\infty}\sum_{k=1}^{n} \mathcal{F}^{-1}\left(|\xi|^\alpha e^{i\xi^3}\varphi(\xi 2^{-k})\right).$$

It is easy to see that the first summand is a bounded function. The kth term is

$$\int_{\mathbb{R}} e^{i\xi^3 + ix\xi}\varphi(\xi 2^{-k})|\xi|^\alpha d\xi.$$

Therefore, by a change of variable, and denoting $\phi(\xi) = \varphi(\xi)|\xi|^\alpha$, we obtain

$$|D^\alpha A_1(x)| \lesssim 1 + \sum_{k\geq 1} 2^{k(1+\alpha)}\left|\Phi\left(2^{3k}, 2^k x\right)\right|,$$

where Φ is as in Lemma 2.2. Using the bounds in that lemma, we obtain (for $0 \leq \alpha \leq \frac{1}{2}$)

$$|D^\alpha A_1(x)| \lesssim 1 + \sum_{k\geq 1:\, 2^{2k}\approx|x|} 2^{k(1+\alpha)}2^{-\frac{3k}{2}} + \sum_{k\geq 1:\, 2^{2k}\not\approx|x|} 2^{k(1+\alpha)}2^{-6k} \lesssim 1.$$

□

By complex interpolation (Theorem 1.3), we now derive:

Theorem 2.3 *(Dispersive decay estimate) For any $\theta \in [0, 1]$ and $\alpha \in [0, 1/2]$, we have*

$$\|D^{\alpha\theta} W_t g\|_{L_x^{2/(1-\theta)}} \lesssim |t|^{-\theta\frac{\alpha+1}{3}}\|g\|_{L^{2/(1+\theta)}}.$$

Proof Proposition 2.1 gives the case $\theta = 1$

$$\|D^\alpha W_t g\|_{L^\infty} \lesssim |t|^{-\frac{\alpha+1}{3}}\|g\|_{L^1}.$$

The theorem will follow from complex interpolation of this bound with the L^2

conservation bound. To do this, consider the analytic family of operators

$$\begin{aligned} T_z g &= |t|^{z\frac{\alpha+1}{3}} D^{\alpha z} W_t\, g \\ &= |t|^{z\frac{\alpha+1}{3}} D^{\alpha z} A_t * g, \end{aligned}$$

where $z = x + iy$, $x \in [0,1]$, $y \in \mathbb{R}$. Since $D^{i\alpha y}$ is unitary, the operator is uniformly bounded in L^2 for $x = 0$. Repeating the proof of Proposition 2.1 with $|\xi|^{\alpha(1+iy)}$ instead of $|\xi|^{\alpha}$ gives

$$\|T_{1+iy} g\|_{L^\infty} \lesssim \langle y \rangle^2 \|g\|_{L^1}.$$

Therefore, complex interpolation between the lines $\Re(z) = 0$ and $\Re(z) = 1$ yields the theorem. □

Theorem 2.4 *(Strichartz estimates) For any $\theta \in [0,1]$ and $\alpha \in [0,1/2]$, we have*

$$\|D^{\alpha\theta/2} W_t\, g\|_{L_t^q L_x^r} \lesssim \|g\|_{L^2} \tag{2.6}$$

$$\left\| \int_0^t D^{\alpha\theta} W_{t-\tau} F(\cdot,\tau) d\tau \right\|_{L_t^q L_x^r} \lesssim \|F\|_{L_t^{q'} L_x^{r'}}, \tag{2.7}$$

where $(q,r) = \left(\frac{6}{\theta(\alpha+1)}, \frac{2}{1-\theta}\right)$.

Proof First note that W_t commutes with D^{α}. We apply the TT^* argument with

$$Tg(x,t) = D^{\alpha\theta/2} W_t g(x),$$

$$T^* F(x) = \int_{\mathbb{R}} D^{\alpha\theta/2} W_{-\tau} F(x,\tau) d\tau.$$

Thus, (2.6) follows from the bound

$$\begin{aligned} \|TT^* F\|_{L_t^q L_x^r} &= \left\| \int_{\mathbb{R}} D^{\alpha\theta} W_{t-\tau} F(\cdot,\tau) d\tau \right\|_{L_t^q L_x^r} \\ &\leq \left\| \int_{\mathbb{R}} \left\| D^{\alpha\theta} W_{t-\tau} F(\cdot,\tau) \right\|_{L_x^r} d\tau \right\|_{L_t^q} \\ &\lesssim \left\| \int_{\mathbb{R}} |t-\tau|^{-\theta(\alpha+1)/3} \|F\|_{L_x^{r'}} d\tau \right\|_{L_t^q} \lesssim \|F\|_{L_t^{q'} L_x^{r'}}. \end{aligned}$$

Here, we used the Minkowski integral inequality, the dispersive bound above, and the Hardy–Littlewood–Sobolev inequality in that order. The inequality

(2.7) is proved similarly

$$\Big\| \int_0^t D^{\alpha\theta} W_{t-\tau} F(\cdot,\tau) d\tau \Big\|_{L_t^q L_x^r}$$

$$\leq \Big\| \int_0^t \|D^{\alpha\theta} W_{t-\tau} F(\cdot,\tau)\|_{L_x^r} d\tau \Big\|_{L_t^q} \lesssim \Big\| \int_0^t |t-\tau|^{-\theta(\alpha+1)/3} \|F\|_{L_x^{r'}} d\tau \Big\|_{L_t^q}$$

$$\leq \Big\| \int_{\mathbb{R}} |t-\tau|^{-\theta(\alpha+1)/3} \|F\|_{L_x^{r'}} d\tau \Big\|_{L_t^q} \lesssim \|F\|_{L_t^{q'} L_x^{r'}}.$$

□

Note that in particular we have the bounds

$$\|D^{\frac{1}{4}} W_t g\|_{L_t^4 L_x^\infty} \lesssim \|g\|_{L^2}, \tag{2.8}$$

$$\|W_t g\|_{L_t^8 L_x^8} \lesssim \|g\|_{L^2}, \tag{2.9}$$

for $(\alpha,\theta) = (\frac{1}{2}, 1)$ and $(\alpha,\theta) = (0, \frac{3}{4})$ respectively.

The derivative gain in Theorem 2.4 is not enough to overcome the derivative nonlinearity in the KdV equation. Nevertheless, by changing the order of the mixed norms one can gain a full derivative:

Theorem 2.5 *(Kato smoothing)*

$$\|\partial_x W_t g\|_{L_x^\infty L_t^2} \lesssim \|g\|_{L^2}.$$

See Exercise 2.3 and Exercise 2.4 for Kato smoothing type estimates for the linear Schrödinger's equation on $\mathbb{R}$.

Proof Writing

$$\partial_x W_t g = \frac{i}{\sqrt{2\pi}} \int_{\mathbb{R}} \xi e^{i\xi^3 t + i\xi x} \widehat{g}(\xi)\, d\xi$$

$$\overset{\eta=\xi^3}{=} \frac{i}{3\sqrt{2\pi}} \int_{\mathbb{R}} \eta^{-1/3} e^{i\eta t + i\eta^{1/3} x} \widehat{g}(\eta^{1/3})\, d\eta,$$

we see that (by Plancherel)

$$\|\partial_x W_t g\|_{L_t^2} = \frac{1}{3} \big\| \eta^{-1/3} e^{i\eta^{1/3} x} \widehat{g}(\eta^{1/3}) \big\|_{L_\eta^2} \overset{\xi=\eta^{1/3}}{=} \|\widehat{g}\|_{L^2} = \|g\|_{L^2}.$$

□

Since the Kato smoothing estimate loses integrability in x, to close the argument in the proof of the local wellposedness of the KdV equation one needs the following maximal function estimate for the Airy evolution; see Kenig–Ponce–Vega [88]:

Theorem 2.6 *For any $s > 3/4$*

$$\|W_t g\|_{L^2_x L^\infty_{t\in[-T,T]}} \lesssim \langle T\rangle^{\frac{1}{2}} \|g\|_{H^s}.$$

We need the following lemma to prove Theorem 2.6:

Lemma 2.7 *Fix a smooth cutoff function ϕ supported in the annulus $\{\xi \in \mathbb{R} : \frac{1}{2} \le |\xi| \le 2\}$ and a smooth cutoff φ supported in $[-2,2]$. For $k \in \mathbb{N}$, let*

$$H_k(x) = \sup_{t\in[-2T,2T]} \left| \int_{\mathbb{R}} e^{i\xi^3 t + ix\xi} \phi(\xi 2^{-k})\, d\xi \right|,$$

and let

$$H_{<1}(x) = \sup_{t\in[-2T,2T]} \left| \int_{\mathbb{R}} e^{i\xi^3 t + ix\xi} \varphi(\xi)\, d\xi \right|.$$

Then for each $k \in \mathbb{N}$, we have

$$\|H_k\|_{L^1} \le C_\phi \langle T\rangle^{\frac{1}{2}} 2^{\frac{3k}{2}},$$

and

$$\|H_{<1}\|_{L^1} \lesssim \langle T\rangle.$$

Proof By a change of variable, we have

$$\begin{aligned} 2^{-k} H_k(2^{-k}x) &= \sup_{t\in[-2T,2T]} \left| \int_{\mathbb{R}} e^{i\xi^3 2^{3k} t + ix\xi} \phi(\xi)\, d\xi \right| \\ &= \sup_{t\in[-T2^{3k+1}, T2^{3k+1}]} |\Phi(t,x)|, \end{aligned}$$

where Φ is as in Lemma 2.2. Therefore, it suffices to prove that for $T > 1$

$$\left\| \sup_{t\in[-T,T]} |\Phi(t,x)| \right\|_{L^1_x} \lesssim T^{\frac{1}{2}}.$$

Using the bounds in Lemma 2.2, we have

$$\sup_{t\in[-T,T]} |\Phi(t,x)| \lesssim \begin{cases} 1, & |x| \le 1, \\ |x|^{-\frac{1}{2}}, & 1 \le |x| \lesssim T, \\ x^{-2}, & |x| \gg T, \end{cases}$$

which yields the claim for H_k.

The claim for $H_{<1}$ is simpler by noting that $|H_{<1}(x)| \lesssim 1$ for each x and $|H_{<1}(x)| \lesssim |x|^{-2}$ for $|x| \gg T$ (by integrating by parts twice as in the proof of Lemma 2.2). □

Proof of Theorem 2.6 Since $s > \frac{3}{4}$, by (1.5) it suffices to prove that

$$\|W_t P_k g\|_{L^2_x L^\infty_{t\in[-T,T]}} \lesssim \langle T\rangle^{\frac{1}{2}} 2^{\frac{3k}{4}} \|g\|_{L^2}$$

for each $k \geq 1$ and that

$$\|W_t P_{<1} g\|_{L^2_x L^\infty_{t\in[-T,T]}} \lesssim \langle T\rangle^{\frac{1}{2}} \|g\|_{L^2}.$$

We only present the proof for $T_k g(x,t) := W_t P_k g$; the proof for $W_t P_{<1} g$ is similar. We have

$$T_k^* f(x) = \int_{-T}^{T} P_k W_\tau^* f(\cdot,\tau) d\tau.$$

Since P_k and W_t commute, we obtain

$$T_k T_k^* f(x,t) = W_t P_k \int_{-T}^{T} P_k W_\tau^* f(\cdot,\tau) d\tau = \int_{-T}^{T} P_k^2 W_{t-\tau} f(\cdot,\tau) d\tau.$$

Note that

$$\begin{aligned} \|T_k T_k^* f(x,t)\|_{L^2_x L^\infty_{t\in[-T,T]}} &\leq \left\| \int_{-T}^{T} \sup_{t\in[-T,T]} |P_k^2 W_{t-\tau} f(\cdot,\tau)| d\tau \right\|_{L^2_x} \\ &\leq \left\| \int_{-T}^{T} H_k * |f(\cdot,\tau)| d\tau \right\|_{L^2_x} = \left\| H_k * \|f(\cdot,t)\|_{L^1_t} \right\|_{L^2_x} \\ &\leq \|H_k\|_{L^1} \|f\|_{L^2_x L^1_t}, \end{aligned}$$

where H_k is as in Lemma 2.7 for a suitable ϕ. By the TT^* lemma, we have

$$\|T_k\|_{L^2 \to L^2_x L^\infty_{t\in[-T,T]}} \leq \|H_k\|_{L^1}^{\frac{1}{2}},$$

which yields the claim by Lemma 2.7. □

2.2 Estimates on the torus

Unlike the case of the real line, there are no dispersive decay estimates on the torus. Bourgain, in [15, 16], obtained analogous Strichartz estimates by using number theoretic arguments. We first consider the Airy equation on the torus

$$\begin{cases} u_t + u_{xxx} = 0, & t \in \mathbb{R}, \ x \in \mathbb{T}, \\ u(0,\cdot) = g(\cdot) \in H^s(\mathbb{T}). \end{cases} \tag{2.10}$$

As in the real line, by applying the Fourier transform, the solution of (2.10) can be expressed as a multiplier operator of the form

$$u(t,x) = W_t g(x) = \sum_{k\in\mathbb{Z}} e^{ikx} e^{ik^3 t} \widehat{g}(k).$$

As above, all H^s norms are conserved. The conservation law and Sobolev embedding imply the following space-time estimates

$$\|W_t g\|_{L_t^\infty L_x^2} \leq \|g\|_{L^2},$$

$$\|W_t g\|_{L_t^\infty L_x^\infty} \lesssim \|g\|_{H^{\frac{1}{2}+}}.$$

In [16], Bourgain considered Strichartz estimates for the Airy evolution (2.10) of the form

$$\|W_t g\|_{L^p_{x,t\in\mathbb{T}}} \lesssim \|g\|_{H^s}, \quad s \geq 0, \quad p \geq 2. \tag{2.11}$$

We require $s \geq 0$, since the L^p norm is bounded below by the L^2 norm, which is equal to the L^2 norm of the initial data. Consider the data

$$g = \sum_{n=1}^{N} e^{inx}.$$

Note that $\|g\|_{H^s} \approx N^{s+\frac{1}{2}}$. Also note that $|W_t g(x)| \gtrsim N$ on the set

$$\left\{(x,t) : |x| \ll \frac{1}{N}, |t| \ll \frac{1}{N^3}\right\},$$

which implies that

$$\|W_t g\|_{L^p_{x,t\in\mathbb{T}}} \gtrsim N^{1-\frac{4}{p}}.$$

Therefore, for $p \geq 8$, one also requires $s \geq \frac{1}{2} - \frac{4}{p}$. In fact, in [16], Bourgain provided an additional counterexample based on more careful estimates on the Weyl sums, proving that $s = 0$ is not sufficient for $p = 8$. Thus, the conjecture is

$$\|W_t g\|_{L^p_{x,t\in\mathbb{T}}} \lesssim \|g\|_{L^2}, \quad 2 < p < 8,$$
$$\|W_t g\|_{L^8_{x,t\in\mathbb{T}}} \lesssim \|g\|_{H^s}, \quad s > 0,$$
$$\|W_t g\|_{L^p_{x,t\in\mathbb{T}}} \lesssim \|g\|_{H^s}, \quad p > 8, \quad s \geq \frac{1}{2} - \frac{4}{p}.$$

This conjecture is known to be true for $p \leq 4$; see Zygmund [159]. In [16], Bourgain obtained the essentially optimal estimate (with an ϵ derivative loss) for $p \in (4,6]$. More recently, the following essentially optimal estimate was obtained by Hu–Li [75]:

Theorem 2.8 *For any $s > \frac{3}{14}$, we have the estimate*

$$\|W_t g\|_{L^{14}_{x,t\in\mathbb{T}}} \lesssim \|g\|_{H^s}.$$

By interpolating with the trivial L^∞ estimate, this validates the conjecture for $p \geq 14$ with an ϵ derivative loss. We present the result from [16]:

Theorem 2.9 *We have the estimates*

$$(i)\ \|W_t g\|_{L^4_{x,t\in\mathbb{T}}} \lesssim \|g\|_{L^2},$$

$$(ii)\ \|W_t g\|_{L^6_{x,t\in\mathbb{T}}} \lesssim \|g\|_{H^\epsilon}, \quad \text{for any } \epsilon > 0.$$

Proof In this proof, one can work on the Fourier side since the exponents are even integers. This will allow us to use simple counting arguments to establish the estimates. We only prove (ii). The proof of (i) is simpler (see the remark below), and it was obtained in [159].

First assume that $\widehat{g} = 0$ outside $[-N, N]$. We write

$$\|W_t g\|^6_{L^6_{x,t\in\mathbb{T}}} = \frac{1}{(2\pi)^6} \sum_{\substack{k_1,k_2,k_3\in[-N,N]\\ j_1,j_2,j_3\in[-N,N]}} \widehat{g}(k_1)\widehat{g}(k_2)\widehat{g}(k_3)\overline{\widehat{g}(j_1)\widehat{g}(j_2)\widehat{g}(j_3)}$$
$$\times \int_0^{2\pi}\int_0^{2\pi} e^{it(k_1^3+k_2^3+k_3^3-j_1^3-j_2^3-j_3^3)+ix(k_1+k_2+k_3-j_1-j_2-j_3)} dt dx.$$

Performing the integration in x, t, we can restrict the sum to resonant terms

$$\|W_t g\|^6_{L^6_{x,t\in\mathbb{T}}} = \frac{1}{(2\pi)^4} \sum_{\substack{k_1^3+k_2^3+k_3^3=j_1^3+j_2^3+j_3^3\\ k_1+k_2+k_3=j_1+j_2+j_3}} \widehat{g}(k_1)\widehat{g}(k_2)\widehat{g}(k_3)\overline{\widehat{g}(j_1)\widehat{g}(j_2)\widehat{g}(j_3)}$$
$$= \frac{1}{(2\pi)^4} \sum_{p,q} \sum_{\substack{(k_1,k_2,k_3)\in A_{p,q}\\ (j_1,j_2,j_3)\in A_{p,q}}} \widehat{g}(k_1)\widehat{g}(k_2)\widehat{g}(k_3)\overline{\widehat{g}(j_1)\widehat{g}(j_2)\widehat{g}(j_3)}$$
$$= \frac{1}{(2\pi)^4} \sum_{p,q} \left| \sum_{(k_1,k_2,k_3)\in A_{p,q}} \widehat{g}(k_1)\widehat{g}(k_2)\widehat{g}(k_3) \right|^2,$$

where

$$A_{p,q} = \left\{(k_1, k_2, k_3) \in [-N, N]^3 : k_1 + k_2 + k_3 = p,\ k_1^3 + k_2^3 + k_3^3 = q\right\}.$$

We claim that for any $\epsilon > 0$, $\#A_{p,q} \lesssim N^\epsilon$. Indeed, writing

$$q - p^3 = k_1^3 + k_2^3 + k_3^3 - (k_1 + k_2 + k_3)^3$$
$$= -3(k_1 + k_2)(k_2 + k_3)(k_1 + k_3),$$

we see that quantities $k_i + k_j$ can take at most N^ϵ different values by using the standard fact that the number of divisors of an integer N is at most N^ϵ for any $\epsilon > 0$; see Exercise 2.5. Since the quantities $k_i + k_j$ uniquely determine k_1, k_2, k_3, we are done.

Using the claim and Cauchy–Schwarz inequality, we see that

$$\begin{aligned}
\|W_t g\|^6_{L^6_{x,t\in\mathbb{T}}} &\lesssim \sum_{p,q} \sum_{(k_1,k_2,k_3)\in A_{p,q}} \left|\widehat{g}(k_1)\widehat{g}(k_2)\widehat{g}(k_3)\right|^2 \sum_{(k'_1,k'_2,k'_3)\in A_{p,q}} 1 \\
&\lesssim N^\epsilon \sum_{p,q} \sum_{(k_1,k_2,k_3)\in A_{p,q}} \left|\widehat{g}(k_1)\widehat{g}(k_2)\widehat{g}(k_3)\right|^2 \\
&= N^\epsilon \sum_{(k_1,k_2,k_3)\in[-N,N]^3} \left|\widehat{g}(k_1)\widehat{g}(k_2)\widehat{g}(k_3)\right|^2 \\
&= N^\epsilon \|g\|^6_{L^2}.
\end{aligned}$$

For arbitrary $g \in H^\epsilon$, we write

$$g = \widehat{g}(0) + \sum_{n\neq 0} g_n,$$

where

$$\widehat{g_n}(j) = \chi_{[2^n,2^{n+1})}(|j|)\widehat{g}(j).$$

Taking the $L^6_{\mathbb{T}^2}$ norm of $W_t g$ and using the inequality above yields the claim.

To estimate the L^4 norm one needs to consider the set $A_{p,q}$ of the form

$$A_{p,q} = \left\{(k_1, k_2) \in [-N, N]^2 : k_1 + k_2 = p, k_1^3 + k_2^3 = q\right\}.$$

Note that this set has cardinality at most 4. □

In the case of the linear Schrödinger equation on the torus, Bourgain obtained essentially optimal estimates for each $p > 2$ in [15], see Theorem 2.10 below. For higher dimensional tori, the optimal estimates were obtained by Bourgain–Demeter in [22], and they are much harder to prove.

Theorem 2.10 *We have the estimates*

$$i)\ \|e^{it\partial_{xx}} g\|_{L^4_{x,t\in\mathbb{T}}} \lesssim \|g\|_{L^2},$$

$$ii)\ \|e^{it\partial_{xx}} g\|_{L^6_{x,t\in\mathbb{T}}} \lesssim \|g\|_{H^\epsilon},\quad \textit{for any } \epsilon > 0.$$

Proof As above the proof follows from estimates on the cardinality of the sets $A_{p,q}$. For the L^6 bound these sets are

$$\begin{aligned}
A_{p,q} &= \left\{(k_1, k_2, k_3) \in [-N, N]^3 : k_1 + k_2 + k_3 = p, k_1^2 + k_2^2 + k_3^2 = q\right\} \\
&= \left\{(k_1, k_2) \in [-N, N]^2 : k_1^2 + k_2^2 + (p - k_1 - k_2)^2 = q\right\}. \qquad (2.12)
\end{aligned}$$

Note that the cardinality of this set is N^ϵ because of a bound on the number of lattice points on an ellipse, see Bombieri–Pila [12, Theorem 3], Bourgain [15], or Lemma 2.11 below.

It is also easy to see that the cardinality of the set $A_{p,q}$ for the L^4 bound is at most 2. □

We now prove the following lemma:

Lemma 2.11 *For any $\epsilon > 0$, the cardinality of the set $A_{p,q}$ defined in* (2.12) *is $\lesssim N^\epsilon$.*

Proof Note that we can rewrite the set $A_{p,q}$ as

$$A_{p,q} = \left\{(k_1, k_2) \in [-N, N]^2 : (3k_2 - p)^2 + 3(2k_1 + k_2 - p)^2 = 6q - 2p^2\right\}.$$

Since we can assume that $|q| \lesssim N^2$ and $|p| \lesssim N$, and since for given p, q, the quantities

$$X := 3k_2 - p, \quad Y := 2k_1 + k_2 - p$$

uniquely determine (k_1, k_2), it suffices to prove that the equation

$$X^2 + 3Y^2 = \left(X + i\sqrt{3}Y\right)\left(X - i\sqrt{3}Y\right) = A$$

has at most A^ϵ integer solutions. Note that $X \pm i\sqrt{3}Y \in \mathbb{Z}[w]$, where $w = e^{\frac{2\pi i}{3}}$. Therefore, it suffices to prove that A has at most A^ϵ divisors in $Z[w]$.

Since $1+w+w^2 = 0$, we have $\mathbb{Z}[w] = \mathbb{Z}+w\mathbb{Z}$. Recall that $Z[w]$ is a Euclidean domain with the norm

$$N(a + wb) = |a + bw|^2 = a^2 - ab + b^2.$$

There are only 6 units in $Z[w]$, which are given by

$$(a, b) = (\pm 1, \pm 1), (\pm 1, 0), (0, \pm 1).$$

Since every Euclidean domain is a unique factorization domain, we have a representation

$$A = p_1^{a_1} \cdots p_k^{a_k},$$

where $p_j \in Z[w]$ are primes unique up to multiplication by units, and the a_js are natural numbers; see, e.g., Hungerford [76]. The proof now proceeds as in Exercise 2.5 noting that

$$N(A) = N(p_1)^{a_1} \cdots N(p_k)^{a_k},$$

and $N(p) \geq 2$ for any prime p (in fact for any nonzero member of $\mathbb{Z}(w)$ which is not a unit). Also note that for any K, there are only finitely many members of $Z[w]$ with the property $N(a + bw) \leq K$, since $N(a + wb) = |a + bw|^2$. □

We finish this chapter by establishing Strichartz estimates for the linear fractional Schrödinger equation on the torus

$$\begin{cases} iu_t - D^{2\alpha}u = 0, & t \in \mathbb{R},\ x \in \mathbb{T},\ \alpha \in (1/2, 1), \\ u(0, \cdot) = g(\cdot) \in H^s(\mathbb{T}). \end{cases} \tag{2.13}$$

As before, the solution $e^{itD^{2\alpha}}$ can be expressed via the Fourier transform

$$e^{itD^{2\alpha}} = \mathcal{F}^{-1}\left(e^{itk^{2\alpha}}\widehat{g}(k)\right).$$

Theorem 2.12 *[51] For any $\alpha \in \left(\frac{1}{2}, 1\right)$ and $s > \frac{1-\alpha}{4}$, we have*

$$\left\|e^{itD^{2\alpha}}g\right\|_{L^4_{x,t\in\mathbb{T}}} \lesssim \|g\|_{H^s}.$$

Unlike the case of the polynomial dispersion, the contribution of the nonresonant terms is not identically zero. To prove the estimate, we rely on an explicit bound on the time integral of the exponential for the nonresonant terms. This requires the following lemma:

Lemma 2.13 *Fix $\alpha \in (1/2, 1]$. For $n, j, k \in \mathbb{Z}$, we have*

$$G(j, k, n) := \left|(n + k)^{2\alpha} - (n + j + k)^{2\alpha} + (n + j)^{2\alpha} - n^{2\alpha}\right| \gtrsim \frac{|k||j|}{(|k| + |j| + |n|)^{2-2\alpha}},$$

where the implicit constant depends on α.

Proof Let $f_c(x) = (x + c)^{2\alpha} - (x - c)^{2\alpha}$. We have

$$G(j, k, n) = \left|f_{\frac{j}{2}}(n + j/2) - f_{\frac{j}{2}}(n + k + j/2)\right|.$$

We claim that

$$f_c'(x) \gtrsim \frac{|c|}{\max(|c|, |x|)^{2-2\alpha}}.$$

Using the claim, we have by the mean value theorem (for $j, k \neq 0$)

$$\begin{aligned} G(j, k, n) &= \left|f_{\frac{j}{2}}(n + j/2) - f_{\frac{j}{2}}(n + k + j/2)\right| \\ &\gtrsim |k||j| \min_{\gamma\in\left(n+\frac{j}{2}, n+k+\frac{j}{2}\right)} \frac{1}{\max\left(\frac{|j|}{2}, |\gamma|\right)^{2-2\alpha}} \\ &\gtrsim \frac{|k||j|}{(|k| + |j| + |n|)^{2-2\alpha}}. \end{aligned}$$

It remains to prove the claim. Since f_c is odd, and $j \neq 0$, it suffices to consider $x \geq 0$ and $c \gtrsim 1$. In this case, we have

$$f_c'(x) = 2\alpha\left[(x + c)^{2\alpha-1} + (x - c)^{2(\alpha-1)}(c - x)\right].$$

We consider three cases:
Case 1: $0 \le x \le c$.
We have

$$f_c'(x) = 2\alpha\left[(x+c)^{2\alpha-1} + (c-x)^{2\alpha-1}\right],$$

and hence

$$f_c'(x) \gtrsim c^{2\alpha-1}.$$

Case 2: $c \le x \lesssim c$.
We have

$$f_c'(x) = 2\alpha\left[(x+c)^{2\alpha-1} - (x-c)^{2\alpha-1}\right].$$

Therefore, we get

$$f_c'(x) \gtrsim c^{2\alpha-1}\left(\left(\frac{x}{c}+1\right)^{2\alpha-1} - \left(\frac{x}{c}-1\right)^{2\alpha-1}\right) \gtrsim c^{2\alpha-1}.$$

Case 3: $x \gg c$.
We have

$$f_c'(x) = 2\alpha[(x+c)^{2\alpha-1} - (x-c)^{2\alpha-1}].$$

Thus, we have

$$\begin{aligned} f_c'(x) &= 2\alpha x^{2\alpha-1}\Big((1+\frac{c}{x})^{2\alpha-1} - (1-\frac{c}{x})^{2\alpha-1}\Big) \\ &\approx x^{2\alpha-1}\frac{c}{x} = x^{2\alpha-2}c. \end{aligned}$$

Hence, in all cases we have

$$f_c'(x) \gtrsim \frac{|c|}{\max(|c|,|x|)^{2-2\alpha}}.$$

□

Proof of Theorem 2.12 Notice that by the Sobolev embedding theorem we can always take $s < \frac{1}{4}$ in this proof. Calling $f = J^s g$, and denoting $\widehat{f}(k)$ by f_k, we write

$$\begin{aligned}
&\|e^{itD^{2\alpha}}g\|^4_{L^4_tL^4_x} \\
&= \frac{1}{(2\pi)^4}\int_0^{2\pi}\int_0^{2\pi}\sum_{k_1,k_2,k_3,k_4}\frac{e^{it(k_1^{2\alpha}-k_2^{2\alpha}+k_3^{2\alpha}-k_4^{2\alpha})}e^{ix(k_1-k_2+k_3-k_4)}f_{k_1}\overline{f_{k_2}}f_{k_3}\overline{f_{k_4}}}{\langle k_1\rangle^s\langle k_2\rangle^s\langle k_3\rangle^s\langle k_4\rangle^s}dxdt \\
&= \frac{1}{(2\pi)^3}\int_0^{2\pi}\sum_{k_1-k_2+k_3-k_4=0}\frac{e^{it(k_1^{2\alpha}-k_2^{2\alpha}+k_3^{2\alpha}-k_4^{2\alpha})}f_{k_1}\overline{f_{k_2}}f_{k_3}\overline{f_{k_4}}}{\langle k_1\rangle^s\langle k_2\rangle^s\langle k_3\rangle^s\langle k_4\rangle^s}dt \\
&\lesssim \sum_{k_1-k_2+k_3-k_4=0}\frac{|f_{k_1}||f_{k_2}||f_{k_3}||f_{k_4}|}{\langle k_1\rangle^s\langle k_2\rangle^s\langle k_3\rangle^s\langle k_4\rangle^s}\frac{1}{\max\left(1,\left|k_1^{2\alpha}-k_2^{2\alpha}+k_3^{2\alpha}-k_4^{2\alpha}\right|\right)},
\end{aligned}$$

where we used the bound

$$\int_0^{2\pi}e^{ita}dt = O(\min(1,|a|^{-1})).$$

Renaming the variables as

$$k_1 = n+j,\ \ k_2 = n+k+j,\ \ k_3 = n+k,\ \ \text{and}\ \ k_4 = n,$$

and using Lemma 2.13, we get

$$\|e^{itD^{2\alpha}}g\|^4_{L^4_tL^4_x} \lesssim \sum_{n,k,j}\frac{|f_n||f_{n+j}||f_{n+k}||f_{n+k+j}|}{\langle n\rangle^s\langle n+k\rangle^s\langle n+j\rangle^s\langle n+k+j\rangle^s}\frac{1}{\max\left(1,\frac{|kj|}{(|k|+|j|+|n|)^{2-2\alpha}}\right)}.$$

Consider the resonant terms with

$$|kj| \ll (|k|+|j|+|n|)^{2-2\alpha}.$$

Noting that if $kj \neq 0$, then

$$\begin{aligned}
&|kj| \ll (|k|+|j|+|n|)^{2-2\alpha} \\
&\qquad\qquad \lesssim |k|^{2-2\alpha}+|j|^{2-2\alpha}+|n|^{2-2\alpha} \lesssim |kj|^{2-2\alpha}+|n|^{2-2\alpha}.
\end{aligned}$$

Therefore, $|kj| \lesssim |n|^{2-2\alpha}$. We thus divide the sum above into two sums I, II. The sum I is on the set

$$\left\{(j,k,n) : n \neq 0, 0 < |kj| \lesssim |n|^{2-2\alpha}\right\} \cup \{(j,k,n) : kj = 0\},$$

and II is on the set

$$\left\{(j,k,n) : |kj| \gtrsim |n|^{2-2\alpha}\right\}.$$

We have

$$I \lesssim \sum_{\substack{n,k,j \\ 0<|kj|\leq|n|^{2-2\alpha}}} \frac{|f_n||f_{n+j}||f_{n+k}||f_{n+k+j}|}{\langle n\rangle^s\langle n+k\rangle^s\langle n+j\rangle^s\langle n+k+j\rangle^s} + \sum_{j,n}|f_n|^2|f_{n+j}|^2 + \sum_{k,n}|f_n|^2|f_{n+k}|^2.$$

The last two sums on the right-hand side are equal to $\|f\|_{L^2}^4$. We estimate the first sum using the Cauchy–Schwarz inequality to get

$$\lesssim \left[\sum_{\substack{n,k,j \\ 0<|kj|\leq|n|^{2-2\alpha}}} \frac{|f_n|^2}{\langle n\rangle^{2s}\langle n+k\rangle^{2s}\langle n+j\rangle^{2s}\langle n+k+j\rangle^{2s}}\right]^{1/2} \times \left[\sum_{n,k,j}|f_{n+j}f_{n+k}f_{n+k+j}|^2\right]^{1/2}$$

$$\lesssim \|f\|_{L^2}^4 \sup_n \left[\sum_{\substack{k,j \\ 0<|kj|\leq|n|^{2-2\alpha}}} \frac{1}{\langle n\rangle^{2s}\langle n+k\rangle^{2s}\langle n+j\rangle^{2s}\langle n+k+j\rangle^{2s}}\right]^{1/2}.$$

The condition on the sum implies, except for finitely many n, that $|k| \ll |n|$ and $|j| \ll |n|$. Therefore

$$\sum_{\substack{k,j \\ 0<|kj|\leq|n|^{2-2\alpha}}} \frac{1}{\langle n\rangle^{2s}\langle n+k\rangle^{2s}\langle n+j\rangle^{2s}\langle n+k+j\rangle^{2s}} \lesssim \frac{1}{\langle n\rangle^{8s}} \sum_{0<|kj|\leq|n|^{2-2\alpha}} 1 \lesssim \langle n\rangle^{2-2\alpha-8s}\log\langle n\rangle \lesssim 1$$

provided that $s > \frac{1-\alpha}{4}$.

Now we estimate II

$$II \lesssim \sum_{\substack{n,k,j \\ |kj|\geq|n|^{2-2\alpha}}} \frac{|f_n||f_{n+j}||f_{n+k}||f_{n+k+j}|(|n|+|k|+|j|)^{2-2\alpha}}{\langle n\rangle^s\langle n+k\rangle^s\langle n+j\rangle^s\langle n+k+j\rangle^s|kj|}.$$

Using the symmetry in k and j, we have

$$II \lesssim \sum_{\substack{n,k,j \\ |kj|\geq|n|^{2-2\alpha},\ |k|\geq|j|}} \frac{|f_n||f_{n+j}||f_{n+k}||f_{n+k+j}|(|n|+|k|)^{2-2\alpha}}{\langle n\rangle^s\langle n+k\rangle^s\langle n+j\rangle^s\langle n+k+j\rangle^s|kj|}.$$

To estimate the sum we consider three frequency regions, $|k| \approx |n|$, $|k| \ll |n|$, and $|k| \gg |n|$ (we note that the terms with $n = 0$ are in the third region).

Region 1. $|k| \approx |n|$.

In this region, using the Cauchy–Schwarz inequality as above, it suffices to show that the sum

$$\sum_{\substack{|k|\geq|j| \\ |k|\approx|n|}} \frac{(|n| + |k|)^{4-4\alpha}}{\langle n\rangle^{2s}\langle n+k\rangle^{2s}\langle n+j\rangle^{2s}\langle n+k+j\rangle^{2s}k^2 j^2}$$

is bounded in n. We bound this by

$$\sum_{\substack{|k|\geq|j| \\ |k|\approx|n|}} \frac{|n|^{2-4\alpha-2s}}{\langle n+k\rangle^{2s}\langle n+j\rangle^{2s}\langle n+k+j\rangle^{2s} j^2}.$$

Using the inequality

$$\langle m+j\rangle\langle j\rangle \gtrsim \langle m\rangle,$$

and recalling that $s < \frac{1}{4}$, the sum above is

$$\lesssim \sum_{\substack{|k|\geq|j| \\ |k|\approx|n|}} \frac{|n|^{2-4\alpha-4s}}{\langle n+k\rangle^{4s} j^{2-4s}} \lesssim \langle n\rangle^{2-4\alpha-4s+1-4s}.$$

Here we first summed in j and then in k. The sum is bounded in n provided that $s > \frac{3-4\alpha}{8}$.

Region 2. $|k| \ll |n|$.

As in Region 1, it suffices to show that the sum

$$\sum_{\substack{|j|\leq|k|\ll|n| \\ |kj|\gtrsim|n|^{2-2\alpha}}} \frac{|n|^{4-4\alpha}}{\langle n\rangle^{2s}\langle n+k\rangle^{2s}\langle n+j\rangle^{2s}\langle n+k+j\rangle^{2s}k^2 j^2} \approx \sum_{\substack{|j|\leq|k|\ll|n| \\ |kj|\gtrsim|n|^{2-2\alpha}}} \frac{|n|^{4-4\alpha-8s}}{k^2 j^2}$$

is bounded in n. To this end, notice that

$$\sum_{\substack{|j|\leq|k|\ll|n| \\ |kj|\gtrsim|n|^{2-2\alpha}}} \frac{|n|^{4-4\alpha-8s}}{k^2 j^2} \lesssim \sum_{|j|\leq|k|\ll|n|} \frac{|n|^{4-4\alpha-8s}}{|j||k|\langle n\rangle^{2-2\alpha}} \lesssim \sup_n |n|^{2-2\alpha-8s}\log(|n|)^2,$$

which is finite provided that $s > \frac{1-\alpha}{4}$.

Region 3. $|k| \gg |n|$.

In this region, we bound the sum by the Cauchy–Schwarz inequality as follows

$$\sum_{\substack{|j|\le|k|,\,|n|\ll|k|\\|kj|\gtrsim|n|^{2-2\alpha}}} \frac{|f_n||f_{n+j}||f_{n+k+j}||f_{n+k}||k|^{2-2\alpha}}{\langle n\rangle^s\langle n+k\rangle^s\langle n+j\rangle^s\langle n+k+j\rangle^s|kj|}$$

$$\lesssim \Bigg(\sum_{|j|\le|k|,\,|n|\ll|k|} \frac{|f_{n+k}|^2|k|^{4-4\alpha}}{\langle n\rangle^{2s}\langle n+k\rangle^{2s}\langle n+j\rangle^{2s}\langle n+k+j\rangle^{2s}k^2j^2}\Bigg)^{1/2}$$

$$\times\Bigg(\sum_{n,k,j}|f_n|^2|f_{n+j}|^2|f_{n+k+j}|^2\Bigg)^{1/2}$$

$$\lesssim \|f\|_{L^2}^3\Bigg(\sum_{|j|\le|k|,\,|n|\ll|k|} \frac{|f_{n+k}|^2|k|^{2-4\alpha-2s}}{\langle n\rangle^{2s}\langle n+j\rangle^{2s}\langle n+k+j\rangle^{2s}j^2}\Bigg)^{1/2}.$$

Estimating the j sum in parenthesis as in Region 1, we have

$$\lesssim \sum_{|n|\ll|k|}\frac{|f_{n+k}|^2|k|^{2-4\alpha-2s}}{\langle n\rangle^{4s}\langle n+k\rangle^{2s}} \lesssim \sum_{|n|\ll|k|}\frac{|f_{n+k}|^2|k|^{2-4\alpha-4s}}{\langle n\rangle^{4s}}$$

$$\lesssim \sum_{n,k}|f_{n+k}|^2|k|^{1-2\alpha-4s}\langle n\rangle^{1-2\alpha-4s}.$$

We estimate this using the Cauchy–Schwarz inequality

$$\Bigg[\sum_{n,k}|f_{n+k}|^2|k|^{2-4\alpha-8s}\Bigg]^{\frac{1}{2}}\Bigg[\sum_{n,k}|f_{n+k}|^2\langle n\rangle^{2-4\alpha-8s}\Bigg]^{\frac{1}{2}} \lesssim \|f\|_{L^2}^2,$$

provided that $2-4\alpha-8s<-1$, i.e. $s>\frac{3}{8}-\frac{\alpha}{2}$.

Thus, for $s>\max(\frac{1-\alpha}{4},\frac{3}{8}-\frac{\alpha}{2})=\frac{1-\alpha}{4}$, for $\alpha>\frac{1}{2}$, we obtain the Strichartz estimates. □

2.3 The Talbot effect

In this section, we present a surprising property that the solutions of linear dispersive equations on the torus have in common. In Section 5.3, we will extend some of these results to certain nonlinear dispersive PDEs. The reader may postpone reading this section till after reading Section 5.3, and proceed to the local wellposedness theory.

In a 1836 optical experiment, Talbot [140] observed white light passing through a diffraction grating. He studied the resulting diffraction pattern with the help of a magnifying lens, and noticed that a sharp focused grating pattern reappears at a certain distance, now known as the Talbot distance d. Moreover, at rational multiples of the Talbot distance, overlapping copies of the translated pattern appear instead, with a complexity increasing as the denominator of the rational number increases. He also observed that the diffraction pattern is periodic in distance with period d. Rayleigh [124] calculated the Talbot distance as $d = \frac{\alpha^2}{\lambda}$, where α is the spacing of the grating and λ is the wavelength of the incoming light.

Berry and his collaborators (see, e.g. Berry [6], Berry–Klein [7], Berry–Lewis [8], and Berry–Marzoli–Schleich [9]) studied the Talbot effect in a series of papers. In particular, in [7] Berry and Klein used the linear Schrödinger evolution to model the Talbot effect. They showed that at rational multiples of the Talbot distance, $\frac{p}{q}d$, up to q overlapping copies of the grating pattern reappear. This can be considered as a quantization effect for the linear Schrödinger evolution. In [144], Taylor independently obtained Berry and Klein's quantization result and also extended it to higher dimensional spheres and tori [145]. In particular, he proved that at rational times the solution is a linear combination of finitely many translates of the initial data with the coefficients being Gauss sums. He further showed that some classical identities for Gauss sums can be obtained by an analysis of the linear Schrödinger evolution. We present this quantization result in a more general setting of arbitrary order linear dispersive PDEs on the torus, see Olver [120]

$$\begin{cases} iu_t + P(-i\partial_x)u = 0, & t \in \mathbb{R}, \quad x \in \mathbb{T} = \mathbb{R}/2\pi\mathbb{Z}, \\ u(0,\cdot) = g(\cdot), \end{cases} \tag{2.14}$$

where P is a polynomial with integer coefficients. We have:

Theorem 2.14 *For $t = 2\pi\frac{p}{q}$, $(p,q) = 1$, the solution of (2.14) satisfies*

$$e^{itP(-i\partial_x)}g = \frac{1}{q}\sum_{j=0}^{q-1} G_{p,q}(j)g\left(x - 2\pi\frac{j}{q}\right),$$

where

$$G_{p,q}(j) = \sum_{l=0}^{q-1} e^{2\pi i P(l)\frac{p}{q}} e^{2\pi i l\frac{j}{q}}.$$

Proof This follows from the following distributional quantization identity

$$e^{2\pi i\frac{p}{q}P(-i\partial_x)}\delta = \frac{1}{q}\sum_{j=0}^{q-1} G_{p,q}(j)\delta\left(x - 2\pi\frac{j}{q}\right).$$

To see this, note that

$$e^{2\pi i\frac{p}{q}P(-i\partial_x)}\delta = \frac{1}{2\pi}\sum_{k=-\infty}^{\infty} e^{2\pi i P(k)\frac{p}{q}}e^{ikx}.$$

Writing $k = jq + l$, $l = 0, \ldots, q-1$, and noting that $P(k) = P(l) \pmod q$, we have

$$= \frac{1}{2\pi}\sum_{l=0}^{q-1} e^{2\pi i P(l)\frac{p}{q}}e^{ilx}\sum_{j=-\infty}^{\infty} e^{iqjx}.$$

Using the identity

$$\frac{1}{2\pi}\sum_{j=-\infty}^{\infty} e^{iqjx} = \frac{1}{q}\sum_{j=0}^{q-1}\delta\left(x - 2\pi\frac{j}{q}\right), \tag{2.15}$$

we conclude that

$$\begin{aligned} e^{2\pi i\frac{p}{q}P(-i\partial_x)}\delta &= \frac{1}{q}\sum_{l=0}^{q-1} e^{2\pi i P(l)\frac{p}{q}}e^{ilx}\sum_{j=0}^{q-1}\delta\left(x - 2\pi\frac{j}{q}\right) \\ &= \frac{1}{q}\sum_{j=0}^{q-1}\Big[\sum_{l=0}^{q-1} e^{2\pi i P(l)\frac{p}{q}}e^{2\pi i l\frac{j}{q}}\Big]\delta\left(x - 2\pi\frac{j}{q}\right) \\ &= \frac{1}{q}\sum_{j=0}^{q-1} G_{p,q}(j)g\left(x - 2\pi\frac{j}{q}\right). \quad \square \end{aligned}$$

By Theorem 2.14, for step function initial data g, $e^{itP(-i\partial_x)}g$ is a step function at each $t = 2\pi\frac{p}{q}$, with an increasing complexity in q. This leads to the question of the qualitative properties of the solution at irrational multiplies of 2π. Berry and Klein also addressed this question. In particular, in [7] they observed that at irrational multiples the solution have a fractal nowhere differentiable profile. Meanwhile, in [6] Berry conjectured that for the n-dimensional linear Schrödinger equation confined in a box, the imaginary part $\Im u(x,t)$, the real part $\Re u(x,t)$, and the density $|u(x,t)|^2$ of the solution are fractal sets with dimension $D = n + \frac{1}{2}$ for most irrational times. Recall that the fractal (also known as upper Minkowski) dimension, $\overline{\dim}(E)$, of a bounded subset E of $\mathbb{R}^n$ is given by

$$\limsup_{\epsilon\to 0}\frac{\log(\mathcal{N}(E,\epsilon))}{\log(\frac{1}{\epsilon})},$$

where $\mathcal{N}(E,\epsilon)$ is the minimum number of ϵ-balls required to cover E. Berry also observed that in the one-dimensional case there are space slices whose time fractal dimension is $\frac{7}{4}$ and there are diagonal slices with dimension $\frac{5}{4}$.

The idea that the profile of linear dispersive equations depends on the algebraic properties of time was further exploited by Oskolkov [121], Kapitanski–Rodnianski [78], and Rodnianski [125]. Oskolkov studied a large class of linear dispersive equations on the torus with bounded variation initial data. In the case of the linear Schrödinger and Airy equations, he proved that at irrational multiples of the period the solution is a continuous function. Moreover, if the initial data is also continuous, then the solution is a continuous function of space and time. The proof of this result is involved and it is beyond the scope of this book. Kapitanski–Rodniaski [78] showed that the solution to the linear Schrödinger equation has better regularity properties (measured in Besov spaces) at irrational than at rational times. It is important to note that this subtle effect cannot be observed in the scale of Sobolev spaces since the linear propagator is a unitary operator in Sobolev spaces. In [125], using the result in [78], Rodnianski partially justified Berry's conjecture in one dimension, proving that the dimension of the real and imaginary parts of the linear Schrödinger solution has fractal dimension $\frac{3}{2}$ for almost all t:

Theorem 2.15 *[125] Suppose that $g : \mathbb{T} \to \mathbb{C}$ is of bounded variation, and*

$$g \notin \bigcup_{r>1/2} H^r(\mathbb{T}).$$

Then for almost every t, the dimension of the graph of $\Re e^{it\partial_{xx}} g$ or $\Im e^{it\partial_{xx}} g$ is $\frac{3}{2}$.

In particular, if the initial data is a nonconstant step function, the dimension is $\frac{3}{2}$. Chen and Olver [33, 34], numerically studied the Talbot effect, and showed that this phenomenon persists for more general dispersive equations, both linear and nonlinear. The following theorem from Chousionis–Erdoğan–Tzirakis [36] deals with more general linear dispersive equations. In particular, it gives a proof of some observations made by Chen–Olver [33], [34].

Theorem 2.16 *Let P be a polynomial of degree d with integer coefficients, and $P(0) = 0$. Suppose that $g : \mathbb{T} \to \mathbb{C}$ is of bounded variation. Then for almost every t, $e^{itP(-i\partial_x)} g \in C^\alpha(\mathbb{T})$ for any $0 \le \alpha < 2^{1-d}$. In particular, the dimension of the graph of the real and imaginary parts is at most $2 - 2^{1-d}$.*

Moreover, in the case where P is not an odd polynomial, if in addition $g \notin \bigcup_{\epsilon>0} H^{r_0+\epsilon}$ for some $r_0 \in [\frac{1}{2}, \frac{1}{2} + 2^{-d})$, then for almost all t both the real part and the imaginary part of the graph of $e^{itP(-i\partial_x)} g$ have fractal dimension $D \ge 2 + 2^{1-d} - 2r_0$.

When P is an odd polynomial, the lower bound above holds for the real-valued solutions.

Remark 2.17 For the Airy evolution ($d = 3$) and for $r_0 = \frac{1}{2}$, one has $D \in [\frac{5}{4}, \frac{7}{4}]$.

Before we prove this theorem we need the following lemma.

Lemma 2.18 *Let P be a polynomial, with integer coefficients, which is not odd, and $P(0) = 0$. Let $g : \mathbb{T} \to \mathbb{C}$ be of bounded variation. Assume that*

$$r_0 := \sup\{s : g \in H^s\} \in [1/2, 1).$$

Then, for almost every t, neither the real nor the imaginary parts of $e^{itP(-i\partial_x)}g$ belong to H^r for $r > r_0$.

Proof We prove this for the real part, and the same argument works for the imaginary part. Also, since the Sobolev spaces are nested, we can assume that $r < \frac{r_0+1}{2}$. Note that

$$\Re(e^{itP(-i\partial_x)}g) = \frac{1}{2}\sum_{k\in\mathbb{Z}}\left(e^{itP(k)}\widehat{g}(k) + e^{-itP(-k)}\overline{\widehat{g}(-k)}\right)e^{ikx}.$$

Therefore, it suffices to prove that for a subsequence $\{K_n\}$ of $\mathbb{N}$

$$\sum_{k=1}^{K_n} k^{2r}\left|e^{itP(k)}\widehat{g}(k) + e^{-itP(-k)}\overline{\widehat{g}(-k)}\right|^2 \to \infty \text{ for almost every } t.$$

We have

$$\sum_{k=1}^{K} k^{2r}\left|e^{itP(k)}\widehat{g}(k) + e^{-itP(-k)}\overline{\widehat{g}(-k)}\right|^2 =$$

$$\sum_{k=1}^{K} k^{2r}\left(\left|\widehat{g}(k)\right|^2 + \left|\widehat{g}(-k)\right|^2\right) + 2\Re\left(\sum_{k=1}^{K} k^{2r}e^{it[P(k)+P(-k)]}\widehat{g}(k)\widehat{g}(-k)\right).$$

Since the first sum diverges as $K \to \infty$, it suffices to prove that the second sum converges almost everywhere after passing to a subsequence. As such, it suffices to prove that it converges in $L^2_{t\in\mathbb{T}}$. Note that we can rewrite the series as

$$\sum_{m\in\mathbb{Z}} e^{itm} \sum_{1\le k\le K,\, P(k)+P(-k)=m} k^{2r}\widehat{g}(k)\widehat{g}(-k).$$

Since P is not odd and $P(0) = 0$, every integer in the range of

$$k \to P(k) + P(-k)$$

is attained at most $\deg(P)$ times. Therefore, the L^2 convergence of the series follows by Plancherel as

$$\sum_{k=1}^{\infty} k^{4r} \left|\widehat{g}(k)\right|^2 \left|\widehat{g}(-k)\right|^2 \lesssim \sup_k k^{4r-2r_0-2+} \|g\|_{H^{r_0-}}^2 < \infty.$$

In these inequalities, we used the bound $|\widehat{g}(k)| \lesssim |k|^{-1}$, which follows from the bounded variation assumption, and that $r < \frac{r_0+1}{2}$. □

We now prove Theorem 2.16.

Proof of Theorem 2.16 Consider

$$\begin{aligned} H_{N,t}(x) &= \sum_{0<|k|\le N} \frac{e^{itP(k)+ikx}}{k} \\ &= \sum_{k=1}^{N} \frac{e^{itP(k)+ikx} - e^{itP(-k)-ikx}}{k}. \end{aligned}$$

Let

$$T_{N,t}(x) = \frac{1}{N} \sum_{k=1}^{N} \left[e^{itP(k)+ikx} - e^{itP(-k)-ikx} \right].$$

By the summation by parts formula

$$\sum_{k=1}^{N} f_k(g_{k+1} - g_k) = f_{N+1}g_{N+1} - f_1 g_1 - \sum_{k=1}^{N} g_{k+1}(f_{k+1} - f_k),$$

with $g_k = (k-1)T_{k-1,t}$ and $f_k = \frac{1}{k}$, we have

$$\begin{aligned} H_{N,t}(x) &= \frac{N}{N+1} T_{N,t}(x) + \sum_{k=1}^{N} \frac{T_{k,t}(x)}{k+1} \\ &= T_{N,t}(x) + \sum_{k=1}^{N-1} \frac{T_{k,t}(x)}{k+1}. \end{aligned} \tag{2.16}$$

We need the following well-known result from number theory; see, e.g., Montgomery [113]:

Theorem 2.19 *Let P be a polynomial with integer coefficients of degree $d \ge 2$. Let t satisfy*

$$\left| \frac{t}{2\pi} - \frac{a}{q} \right| \le \frac{1}{q^2} \tag{2.17}$$

for some integers a and q. Then for any $\epsilon > 0$, we have

$$\sup_x \left| \sum_{k=1}^{N} e^{itP(k)+ikx} \right| \lesssim N^{1+\epsilon} \left(\frac{1}{q} + \frac{1}{N} + \frac{q}{N^d} \right)^{2^{1-d}}.$$

Recall that the Dirichlet theorem implies that for every irrational $\frac{t}{2\pi}$, the inequality (2.17) holds for infinitely many integers a, q. Given irrational $\frac{t}{2\pi}$, let $\{q_n\}$ be the increasing sequence of positive qs for which (2.17) holds for some a. The next theorem by Khinchin [94] and Lévy [102] gives quantitative information on the sequence $\{q_n\}$:

Theorem 2.20 *For almost every t, we have*

$$\lim_{n\to\infty} q_n^{1/n} = \gamma,$$

for some absolute constant γ independent of t.

An immediate corollary, see Exercise 2.7, of this theorem is the following:

Corollary 2.21 *For almost every t, and for any $\epsilon > 0$, we have*

$$q_{n+1} \le q_n^{1+\epsilon}$$

for all sufficiently large n.

This in turn implies:

Corollary 2.22 *For almost every t, for any $\epsilon > 0$, and for all sufficiently large N, there exists $q \in [N, N^{1+\epsilon}]$ so that* (2.17) *holds for q and for some a.*

Combining Theorem 2.19 and Corollary 2.22, we obtain:

Corollary 2.23 *For almost every t, and for any $\epsilon > 0$, we have*

$$\sup_x \left| \sum_{k=1}^{N} e^{itP(k)+ikx} \right| \lesssim N^{1-2^{1-d}+\epsilon}$$

for all N.

Using Corollary 2.23, we see that

$$\|T_{N,t}\|_{L^\infty} \lesssim N^{-2^{1-d}+\epsilon}.$$

This together with (2.16) imply that for almost every t the sequence $H_{N,t}$ converges uniformly to a continuous function

$$H_t(x) = \sum_{k\neq 0} \frac{e^{itP(k)+ikx}}{k}.$$

Using the summation by parts formula, it also implies that for any $\epsilon > 0$, and for any $j = 1, 2, \ldots$

$$\left\| \sum_{2^{j-1}\le|k|<2^j} \frac{e^{itP(k)+ikx}}{k} \right\|_{L_x^\infty} \lesssim 2^{-j(2^{1-d}-\epsilon)}.$$

Recall that the Besov space $B^s_{p,\infty}$ is defined via the norm

$$\|f\|_{B^s_{p,\infty}} := \sup_{j\geq 0} 2^{sj}\|P_j f\|_{L^p},$$

where P_j is a Littlewood–Paley projection on to the frequencies $\approx 2^j$. Therefore, for almost every t

$$H_t \in \bigcap_{\epsilon>0} B^{2^{1-d}-\epsilon}_{\infty,\infty}(\mathbb{T}).$$

We should note that $C^\alpha(\mathbb{T})$ coincides with $B^\alpha_{\infty,\infty}(\mathbb{T})$ for $0 < \alpha < 1$; see, e.g., Tribel [149], and Exercise 2.8. Therefore, for almost all t, $H_t \in C^\alpha(\mathbb{T})$ for any $\alpha < 2^{1-d}$.

Now, given function g of bounded variation, we write

$$e^{itP(-i\partial_x)}g = \widehat{g}(0) + \sum_{k\neq 0} e^{itP(k)+ikx}\widehat{g}(k).$$

Moreover, we can write

$$\begin{aligned}\widehat{g}(k) &= \frac{1}{2\pi}\int_{\mathbb{T}} e^{-iky}g(y)dy \\ &= \frac{1}{2\pi ik}\int_{\mathbb{T}} e^{-iky}dg(y),\end{aligned}$$

where dg is the Lebesgue–Stieltjes measure associated with g. Therefore

$$\begin{aligned}e^{itP(-i\partial_x)}g &= \widehat{g}(0) - i\lim_{N\to\infty} H_{N,t} * dg \\ &= \widehat{g}(0) - iH_t * dg\end{aligned}$$

by the uniform convergence of the sequence $H_{N,t}$. In particular, for almost every t,

$$e^{itP(-i\partial_x)}g \in \bigcap_{\epsilon>0} C^{2^{1-d}-\epsilon}(\mathbb{T}) = \bigcap_{\epsilon>0} B^{2^{1-d}-\epsilon}_{\infty,\infty}(\mathbb{T}), \tag{2.18}$$

since the convolution of a finite measure with a C^α function is C^α. Also recall that if $f : \mathbb{T} \to \mathbb{R}$ is in C^α, then the graph of f has fractal dimension $D \leq 2-\alpha$; see Exercise 2.9. Therefore, the graphs of $\Re(u)$ and $\Im(u)$ have dimension at most $2 - 2^{1-d}$ for almost all t.

Now in addition assume that $g \notin H^r(\mathbb{T})$ for any $r > r_0 \geq \frac{1}{2}$. This implies using Lemma 2.18 (considering only the real-valued solution when P is odd) that

$$\Im e^{it\partial_{xx}}g,\ \Re e^{it\partial_{xx}}g \notin \bigcup_{\epsilon>0} B^{2r_0-2^{1-d}+\epsilon}_{1,\infty}(\mathbb{T}), \tag{2.19}$$

since

$$H^r(\mathbb{T}) \supset B^{r_1}_{1,\infty}(\mathbb{T}) \cap B^{r_2}_{\infty,\infty}(\mathbb{T}), \tag{2.20}$$

for $r_1 + r_2 > 2r$; see Exercise 2.10.

The lower bound for the fractal dimension in Theorem 2.16 follows from (2.19) and Theorem 2.24 below. □

The following theorem was proved by Deliu–Jawerth [49]; we give a proof for completeness.

Theorem 2.24 *The graph of a continuous function $f : \mathbb{T} \to \mathbb{R}$ has fractal dimension $D \geq 2 - s$ provided that $f \notin \bigcup_{\epsilon>0} B^{s+\epsilon}_{1,\infty}$.*

Proof It suffices to prove that if f is continuous and if for some $0 < \alpha < 1$

$$\sup_{j\geq 1} 2^{\alpha j} \|P_j f\|_{L^1} = \infty,$$

then the graph E of f has dimension $D \geq 2 - \alpha$.

Since $P_j 1 = 0$ and by Exercise 1.12 and Exercise 2.11, we have

$$\begin{aligned}
\|P_j f\|_{L^1(\mathbb{T})} &= \left\| \int_{\mathbb{T}} \varphi_j(y)(f(x-y) - f(x))dy \right\|_{L^1(\mathbb{T})} \\
&\lesssim \int_{|y|<1/2} \frac{2^j}{(1+2^j|y|)^2} \|f(x-y) - f(x)\|_{L^1_{x\in\mathbb{T}}} dy \\
&\lesssim \int_{|y|<1/2} \frac{2^j y^2}{(1+2^j|y|)^2} \mathcal{N}(E,|y|)dy \\
&\lesssim \sum_{k=1}^{\infty} \int_{|y|\approx 2^{-k}} \frac{2^j y^2}{(1+2^j|y|)^2} \mathcal{N}(E,|y|)dy \\
&\lesssim \sum_{k=1}^{\infty} \frac{2^j 2^{-3k}}{(1+2^j 2^{-k})^2} \mathcal{N}(E,2^{-k}).
\end{aligned}$$

Therefore

$$\sup_{j\geq 1} 2^{\alpha j} \|P_j f\|_{L^1} \lesssim \sum_{k=1}^{\infty} 2^{-3k} \mathcal{N}(E,2^{-k}) \sup_{j\geq 1} \frac{2^{j(\alpha+1)}}{(1+2^j 2^{-k})^2}.$$

Since for $0 < \alpha < 1$

$$\sup_{j\geq 1} \frac{2^{j(\alpha+1)}}{(1+2^j 2^{-k})^2} \approx 2^{k(\alpha+1)},$$

we conclude that

$$\sum_{k=1}^{\infty} 2^{(\alpha-2)k} \mathcal{N}(E,2^{-k})$$

diverges. This implies that

$$\mathcal{N}(E, 2^{-k}) \geq 2^{(2-\alpha)k} k^{-2}$$

for infinitely many k, which yields the claim. □

We now discuss the fractal dimension of the density function $|e^{it\partial_{xx}} g|^2$. This problem was partially resolved in [36] for a large class of step function initial data. It would be interesting to have an analogous result for bounded variation data and more general dispersion as in Theorem 2.16.

Theorem 2.25 *Let g be a nonconstant complex valued step function on $\mathbb{T}$ with jumps only at rational multiples of π. Then for almost every t the graphs of $\left|e^{it\partial_{xx}} g\right|^2$ and $\left|e^{it\partial_{xx}} g\right|$ have fractal dimension $\frac{3}{2}$.*

Note that, since $C^\alpha(\mathbb{T})$ is an algebra, the upper bound is a corollary of Theorem 2.16. The lower bound for $|e^{it\partial_{xx}} g|^2$ yields the lower bound for $\left|e^{it\partial_{xx}} g\right|$ since $\left|e^{it\partial_{xx}} g\right|$ is a continuous and hence bounded function for almost every t (see Exercise 2.12). The lower bound for $\left|e^{it\partial_{xx}} g\right|^2$ follows from the proof of Theorem 2.16 provided that we have

$$\left|e^{it\partial_{xx}} g\right|^2 \in \bigcap_{\epsilon>0} B^{\frac{1}{2}-\epsilon}_{\infty,\infty}(\mathbb{T}),$$

and

$$\left|e^{it\partial_{xx}} g\right|^2 \notin H^{\frac{1}{2}}.$$

The former follows from (2.18) since $B^\alpha_{\infty,\infty} = C^\alpha$ is an algebra. For the latter, we have:

Proposition 2.26 *Let g be a nonconstant complex valued step function on $\mathbb{T}$ with jumps only at rational multiples of π. Then for every irrational value of $\frac{t}{2\pi}$, we have*

$$\left|e^{it\partial_{xx}} g\right|^2 \notin H^{\frac{1}{2}}.$$

Proof We can write after a translation

$$g = \sum_{\ell=1}^{L} c_\ell \chi_{[a_\ell, a_{\ell+1})},$$

where

$$0 = a_1 < a_2 < \ldots < a_L < a_{L+1} = 2\pi,$$

and c_ℓs are complex numbers satisfying $c_\ell \neq c_{\ell+1}$, $\ell = 1, 2, \ldots, L$ (here $c_{L+1} := c_1$).

Recall that a subset S of $\mathbb{N}$ has positive density if

$$\liminf_{N\to\infty} \frac{\#(\{1,\dots,N\}\cap S)}{N} > 0.$$

It suffices to prove that for a positive density subset S of $\mathbb{N}$ we have

$$\forall k \in S, \quad \left|\mathcal{F}(|e^{it\partial_{xx}} g|^2)(k)\right| \gtrsim \frac{1}{k}. \tag{2.21}$$

First note that for $n \neq 0$

$$\widehat{e^{it\partial_{xx}} g}(n) = \frac{i}{2\pi} \sum_{\ell=1}^{L} c_\ell e^{-itn^2} \frac{e^{-ina_{\ell+1}} - e^{-ina_\ell}}{n}.$$

Let K be a natural number such that

$$Ka_\ell = 0 \text{ (mod } 2\pi) \text{ for each } \ell.$$

For nonzero k divisible by K, we have (using $e^{ika_\ell} = 1$ and $\widehat{g}(k) = 0$)

$$\begin{aligned}
&\mathcal{F}\left(\left|e^{it\partial_{xx}} g\right|^2\right)(k) \\
&= \frac{1}{4\pi^2} \sum_{\ell,m=1}^{L} c_\ell \overline{c_m} \sum_{n\neq 0,k} e^{-itn^2} e^{it(n-k)^2} \frac{\left(e^{-ina_{\ell+1}} - e^{-ina_\ell}\right)\left(e^{ina_{m+1}} - e^{ina_m}\right)}{n(n-k)} \\
&= \frac{e^{itk^2}}{4\pi^2 k} \sum_{\ell,m=1}^{L} c_\ell \overline{c_m} \sum_{n\neq 0,k} \frac{e^{-2itnk}}{n-k} \left(e^{-ina_{\ell+1}} - e^{-ina_\ell}\right)\left(e^{ina_{m+1}} - e^{ina_m}\right) \\
&\quad - \frac{e^{itk^2}}{4\pi^2 k} \sum_{\ell,m=1}^{L} c_\ell \overline{c_m} \sum_{n\neq 0,k} \frac{e^{-2itnk}}{n} \left(e^{-ina_{\ell+1}} - e^{-ina_\ell}\right)\left(e^{ina_{m+1}} - e^{ina_m}\right).
\end{aligned}$$

Changing the variable $n - k \to n$ in the first sum, we obtain

$$\begin{aligned}
&\mathcal{F}\left(\left|e^{it\partial_{xx}} g\right|^2\right)(k) = O(1/k^2) \\
&\quad - \frac{i\sin(k^2 t)}{2\pi^2 k} \sum_{\ell,m=1}^{L} c_\ell \overline{c_m} \sum_{n\neq 0} \frac{e^{-2itnk}}{n} \left(e^{-ina_{\ell+1}} - e^{-ina_\ell}\right)\left(e^{ina_{m+1}} - e^{ina_m}\right).
\end{aligned}$$

Using the formula (see Exercise 2.13)

$$\sum_{n\neq 0} \frac{e^{in\alpha}}{n} = i(\pi - \alpha), \quad 0 < \alpha < 2\pi, \tag{2.22}$$

we have

$$\mathcal{F}\left(\left|e^{it\partial_{xx}}g\right|^2\right)(k) = O(1/k^2)$$
$$-\frac{\sin(k^2t)}{2\pi^2 k}\sum_{\ell,m=1}^{L} c_\ell\overline{c_m}(\lfloor -2kt + a_{m+1} - a_{\ell+1}\rfloor - \lfloor -2kt + a_{m+1} - a_\ell\rfloor)$$
$$+\frac{\sin(k^2t)}{2\pi^2 k}\sum_{\ell,m=1}^{L} c_\ell\overline{c_m}(\lfloor -2kt + a_m - a_{\ell+1}\rfloor - \lfloor -2kt + a_m - a_\ell\rfloor),$$

where

$$\lfloor x\rfloor = x \pmod{2\pi} \in [0, 2\pi).$$

Note that for $0 \le a < b \le 2\pi$, we have

$$\lfloor c - b\rfloor - \lfloor c - a\rfloor = a - b + 2\pi\chi_{[a,b)}(\lfloor c\rfloor).$$

Using this, we have

$$\mathcal{F}\left(\left|e^{it\partial_{xx}}g\right|^2\right)(k) = O(1/k^2)$$
$$-\frac{\sin(k^2t)}{\pi k}\sum_{\ell,m=1}^{L} c_\ell\overline{c_m}(\chi_{[a_\ell,a_{\ell+1})}(\lfloor -2kt + a_{m+1}\rfloor) - \chi_{[a_\ell,a_{\ell+1})}(\lfloor -2kt + a_m\rfloor)).$$

From now on, we restrict ourselves to k so that $\lfloor -2kt\rfloor \in (0, \epsilon)$ for some fixed

$$0 < \epsilon < \min_{\ell\in\{1,\dots,L\}} (a_{\ell+1} - a_\ell),$$

where $a_{L+1} = 2\pi$. We note that for such k

$$\chi_{[a_\ell,a_{\ell+1})}(\lfloor -2kt + a_m\rfloor) = 1 \text{ for } m = \ell,$$

and it is zero otherwise. Therefore

$$\mathcal{F}\left(\left|e^{it\partial_{xx}}g\right|^2\right)(k) = O(1/k^2) + \frac{\sin(k^2t)}{\pi k}\left[\sum_{\ell=1}^{L}|c_\ell|^2 - \sum_{\ell=1}^{L} c_\ell\overline{c_{\ell-1}}\right].$$

Note that we can write the quantity in bracket as

$$C\cdot C - C\cdot\tilde{C},$$

where

$$C = (c_1,\dots,c_L) \text{ and } \tilde{C} = (c_L, c_1,\dots,c_{L-1}).$$

We have $\|\tilde{C}\| = \|C\|$, and since the adjacent c_ℓs are distinct, we also have $\tilde{C} \ne C$. Therefore, the absolute value of the quantity in bracket is nonzero.

Thus, (2.21) holds for

$$S = \left\{k \in \mathbb{N} : K|k, \lfloor -2kt \rfloor \in (0, \epsilon), \lfloor k^2 t \rfloor \in [\pi/4, 3\pi/4]\right\}.$$

This set has positive density for any irrational $\frac{t}{2\pi}$ since the set

$$\left\{\left(\lfloor -2Kjt \rfloor, \left\lfloor K^2 j^2 t \right\rfloor\right) : j \in \mathbb{N}\right\}$$

is uniformly distributed on $\mathbb{T}^2 = \mathbb{T} \times \mathbb{T}$ by the classical Weyl's theorem (see e.g. Theorem 6.4 on page 49 in Kuipers–Niederreiter [98]). □

We close this chapter with a discussion of another manifestation of the Talbot effect. Note that Theorem 2.14 implies that for any polynomial P with integer coefficients, $e^{itP(k)}$ is an $L^r(\mathbb{T})$ multiplier for $t = 2\pi\frac{p}{q}$ and for any r. In the case of the linear Schrödinger evolution, we have the following rational/irrational dichotomy:

Theorem 2.27 *For any $t = 2\pi\frac{p}{q}$ with $(p, q) = 1$, and for any $r \in [1, \infty]$, we have*

$$\left\|e^{it\partial_{xx}}\right\|_{L^r(\mathbb{T}) \to L^r(\mathbb{T})} \approx q^{\left|\frac{1}{2} - \frac{1}{r}\right|}.$$

On the other hand, for almost every t, $e^{it\partial_{xx}}$ is unbounded in $L^r(\mathbb{T})$ unless $r = 2$.

Recall that linear Schrödinger evolution on $\mathbb{R}$ has drastically different properties. In particular, $e^{it\xi^2}$ is not an $L^r(\mathbb{R})$ multiplier for any nonzero t when $r \neq 2$; see Exercise 2.14.

Proof We need the following facts:
(a) The Gauss sum

$$G_{p,q}(j) = \sum_{l=0}^{q-1} e^{2\pi i l^2 \frac{p}{q}} e^{2\pi i l \frac{j}{q}}$$

satisfies (see Exercise 2.15)

$$\left|G_{p,q}(j)\right| \le \sqrt{2q},$$

for each j, and

$$\left|G_{p,q}(j)\right| \ge \sqrt{q},$$

for each even or odd j.
(b) A theorem by Khinchin [94] states that for almost every $t > 0$, there are infinitely many relatively prime pairs of natural numbers p, q such that

$$\left|\frac{t}{2\pi} - \frac{p}{q}\right| \le \frac{1}{q^2 \log(q)}. \tag{2.23}$$

Using the upper bound in (a) and Theorem 2.14, we have ($t = 2\pi\frac{p}{q}$)

$$\|e^{it\partial_{xx}}g\|_{L^1} = \frac{1}{q}\left\|\sum_{j=0}^{q-1} G_{p,q}(j)g\left(x - 2\pi\frac{j}{q}\right)\right\|_{L^1}$$

$$\lesssim \frac{1}{\sqrt{q}}\sum_{j=0}^{q-1}\left\|g\left(x - 2\pi\frac{j}{q}\right)\right\|_{L^1} \lesssim \sqrt{q}\|g\|_{L^1}.$$

Interpolating this with the trivial L^2 bound and duality yields the upper bound for the multiplier norm. The lower bound follows from the lower bound on the Gauss sums by taking g supported in $[0, \frac{1}{q}]$ with L^r norm 1. Indeed, using the support condition and the lower bound above, we have ($t = 2\pi\frac{p}{q}$)

$$\|e^{it\partial_{xx}}g\|_{L^r}^r = \frac{1}{q^r}\left\|\sum_{j=0}^{q-1} G_{p,q}(j)g\left(x - 2\pi\frac{j}{q}\right)\right\|_{L^r}^r$$

$$= \frac{1}{q^r}\sum_{j=0}^{q-1}|G_{p,q}(j)|^r\left\|g\left(x - 2\pi\frac{j}{q}\right)\right\|_{L^r}^r$$

$$\gtrsim \frac{1}{q^r}\sum_{j=0}^{q-1}|G_{p,q}(j)|^r \geq q^{1-\frac{r}{2}}. \quad (2.24)$$

This gives the lower bound for $r \in [1, 2]$, and by duality the lower bound holds also for $r > 2$.

It suffices to prove the second part of the theorem for $r \in [1, 2)$. Fix t so that (2.23) holds for a sequence $(p_n, q_n) = 1$, $q_n \to \infty$. For each n, let

$$g_n(x) = q_n^{\frac{1}{r}}\phi(xq_n),$$

where ϕ is a smooth bump function supported in $[-1, 1]$. Note that $\|g_n\|_r \approx 1$, and by scaling and decay of $\widehat{\phi}$, we have

$$|\widehat{g_n}(k)| \lesssim q_n^{\frac{1}{r}-1}\langle k/q_n\rangle^{-5}.$$

By the calculations in (2.24), we have

$$\left\|e^{i2\pi\frac{p_n}{q_n}\partial_{xx}}g_n\right\|_{L^r} \gtrsim q_n^{\frac{1}{r}-\frac{1}{2}}.$$

On the other hand

$$\left\| e^{it\partial_{xx}} g_n - e^{i2\pi \frac{p_n}{q_n} \partial_{xx}} g_n \right\|_{L^r} \lesssim \left\| e^{it\partial_{xx}} g_n - e^{i2\pi \frac{p_n}{q_n} \partial_{xx}} g_n \right\|_{L^2}$$
$$= \left\| (e^{itk^2} - e^{i2\pi \frac{p_n}{q_n} k^2}) \widehat{g_n}(k) \right\|_{\ell^2} \lesssim \left\| \left| t - 2\pi \frac{p_n}{q_n} \right| k^2 \widehat{g_n}(k) \right\|_{\ell^2}$$
$$\lesssim \frac{1}{q_n^2 \log(q_n)} q_n^{\frac{1}{r}-1} \left\| k^2 \langle k/q_n \rangle^{-5} \right\|_{\ell^2} \lesssim \frac{q_n^{\frac{1}{r}-\frac{1}{2}}}{\log(q_n)},$$

where we used the bound

$$\left\| k^2 \langle k/q_n \rangle^{-5} \right\|_{\ell^2} \approx q_n^{\frac{5}{2}}.$$

Combining these bounds using triangle inequality, we have (for large n)

$$\left\| e^{it\partial_{xx}} g_n \right\|_{L^r} \gtrsim q_n^{\frac{1}{r}-\frac{1}{2}}.$$

This finishes the proof since $\|g_n\|_r \approx 1$ for each n. □

Exercises

2.1 Fix $n \geq 1$. We call a pair (q, r) of exponents admissible if

$$2 \leq q, r \leq \infty, \quad \frac{2}{q} + \frac{n}{r} = \frac{n}{2} \text{ and}$$
$$(q, r, n) \neq (2, \infty, 2).$$

Prove that for any nonendpoint ($q > 2$) admissible exponents (q, r) and $(\tilde{q}, \tilde{r})$ we have the following estimates: The homogeneous estimate

$$\left\| e^{it\Delta} g \right\|_{L_t^q L_x^r(\mathbb{R} \times \mathbb{R}^n)} \lesssim \|g\|_{L^2}, \tag{2.25}$$

and the inhomogeneous estimate

$$\left\| \int_0^t e^{i(t-s)\Delta} F(s) ds \right\|_{L_t^q L_x^r(\mathbb{R} \times \mathbb{R}^n)} \lesssim \|F\|_{L_t^{\tilde{q}'} L_x^{\tilde{r}'}(\mathbb{R} \times \mathbb{R}^n)}, \tag{2.26}$$

where $\frac{1}{\tilde{q}} + \frac{1}{\tilde{q}'} = 1$ and $\frac{1}{\tilde{r}} + \frac{1}{\tilde{r}'} = 1$. See Exercise 1.7 and Exercise 1.8.

2.2 Establish the $L^1 \to L^\infty$ dispersive estimate for the wave equation

$$\left\| e^{\pm itD} g \right\|_{L^\infty(\mathbb{R}^n)} \lesssim \langle t \rangle^{-\frac{n-1}{2}} \|g\|_{L^1(\mathbb{R}^n)}$$

for any $g \in \mathcal{S}(\mathbb{R}^n)$ whose Fourier transform is supported in $\{\xi : 1 \leq |\xi| \leq 2\}$. Note that the decay rate is weaker than the Schrödinger decay rate since the dispersion is weaker. Also note that there is no decay in one-dimension since the solution is a traveling wave.

Hint: Use the decay rate of the Fourier transform of the surface measure σ of the unit sphere S^{n-1} (see Stein [133])

$$\left|\widehat{\sigma}(\xi)\right| \lesssim \langle \xi \rangle^{-\frac{n-1}{2}}, \quad \xi \in \mathbb{R}^n.$$

2.3 Using arguments similar to the proof of Theorem 2.5, prove that

$$\sup_x \int_{-\infty}^{\infty} \left| D^{\frac{1}{2}} e^{it\partial_{xx}} g \right|^2 dt \leq C\|g\|^2_{L^2(\mathbb{R})}.$$

2.4 For $s \geq 0$, prove that

$$\sup_x \left\| \eta(t) e^{it\partial_{xx}} g \right\|_{H_t^{\frac{2s+1}{4}}} \lesssim \|g\|_{H^s},$$

where $\eta \in C_0^\infty(\mathbb{R})$.

2.5 Prove that the number of divisors, $d(N)$, of an integer N is bounded by $C_\epsilon N^\epsilon$ for any $\epsilon > 0$ by following the steps below:
(a) Prove that the number of divisors of $N = p_1^{a_1} p_2^{a_2} \ldots p_k^{a_k}$ is equal to

$$\prod_{j=1}^{k} (a_j + 1).$$

(b) Prove that $\frac{a+1}{p^{\epsilon a}}$ is bounded in $a \in \mathbb{N}$ by a constant depending on ϵ for any prime p.
(c) Prove that $\frac{a+1}{p^{\epsilon a}}$ is bounded by 1 if $p^\epsilon > e$.
(d) Complete the proof by noting that the number of primes less than $e^{1/\epsilon}$ contributing to the product $\frac{d(N)}{N^\epsilon}$ is uniformly bounded in N.

2.6 Prove the distributional identity

$$\frac{1}{2\pi} \sum_{j=-\infty}^{\infty} e^{iqjx} = \frac{1}{q} \sum_{j=0}^{q-1} \delta\left(x - 2\pi \frac{j}{q}\right), \quad x \in \mathbb{R}, \; q \in \mathbb{N}.$$

2.7 Prove Corollary 2.21 and Corollary 2.22 using Theorem 2.20.

2.8 Prove that $C^\alpha(\mathbb{T}) = B^\alpha_{\infty,\infty}(\mathbb{T})$.
Hint: Use Exercise 1.12 in one direction and use the mean value theorem and Exercise 1.11 for the other direction.

2.9 Let $\alpha \in [0, 1]$. Prove that if $f : \mathbb{T} \to \mathbb{R}$ is in C^α, then the graph of f has fractal dimension $D \leq 2 - \alpha$.
Hint: Use C^α norm to find an upper bound for the number of boxes needed to cover the graph.

2.10 Prove that

$$H^r(\mathbb{T}) \supset B^{r_1}_{1,\infty}(\mathbb{T}) \cap B^{r_2}_{\infty,\infty}(\mathbb{T}),$$

for $r_1 + r_2 > 2r$.

2.11 (a) Recall that $\mathcal{N}(E, r)$ denotes the minimum number of balls of radius r needed to cover the set E. Let $\widetilde{\mathcal{N}}(E, 2^j)$ denote the minumum number of dyadic cubes of side length 2^j required to cover E. Prove that

$$\mathcal{N}(E, r) \approx \widetilde{\mathcal{N}}(E, 2^j)$$

for $r \approx 2^j$.

(b) Prove that if f is a continuous function on the torus, then

$$\|f(\cdot - r) - f(\cdot)\|_{L^1_{\mathbb{T}}} \lesssim r^2 \mathcal{N}(E, r),$$

where E is the graph of f.

2.12 Let $f : \mathbb{T} \to [0, \infty)$ be continuous. Prove that the dimension of the graph of f^2 is less than or equal to the dimension of the graph of f.

2.13 Prove the identity

$$\sum_{n \neq 0} \frac{e^{in\alpha}}{n} = i(\pi - \alpha), \quad 0 < \alpha < 2\pi$$

using Fourier series.

2.14 Prove that $e^{it\xi^2}$ is not an $L^r(\mathbb{R})$ multiplier for any nonzero t when $r \neq 2$ by considering the evolution of $e^{-\sigma|x|^2}$ as $\sigma \to 0$ and $\sigma \to \infty$.
Hint: Use Exercise 1.8.

2.15 Consider the Gauss sums

$$G_{p,q}(j) = \sum_{l=0}^{q-1} e^{2\pi i l^2 \frac{p}{q}} e^{2\pi i l \frac{j}{q}}.$$

Prove that

$$|G_{p,q}(j)| \leq \sqrt{2q},$$

for each j, and

$$|G_{p,q}(j)| \geq \sqrt{q},$$

for each even or odd j.
Hint: Consider $|G_{p,q}(j)|^2$ as a double sum in l_1 and l_2. Note that after a change of variable, the phase depends linearly on the summation index.

3

Methods for establishing wellposedness

In this chapter, we present basic methods for establishing the wellposedness theory of nonlinear dispersive PDEs. Various techniques have been developed over the years in order to produce local-in-time solutions in lower regularity spaces. The methods we will consider are the energy method of Bona–Smith [13], the oscillatory integral method of Kenig–Ponce–Vega [88], the restricted norm method of Bourgain [15, 16], and finally the differentiation by parts method of Babin–Ilyin–Titi [1]. Our aim is to present these techniques in complete detail for the model case of the KdV equation. However, these methods are applicable to a wide class of nonlinear dispersive PDEs. In particular, the wellposedness theory of the nonlinear Schrödinger (NLS) equation with a power type nonlinearity is a well-studied subject, and a thorough exposition can be found in Cazenave [30]. We will study the cubic and quintic NLS equations on the torus in Section 3.5 below. For more recent presentations that cover many different nonlinear dispersive models, see Bourgain [20], Sulem–Sulem [137], Tao [143], and Linares–Ponce [105].

Consider a dispersive PDE of the form

$$\begin{cases} iu_t = Lu + F(u), & t \in \mathbb{R}, \quad x \in K = \mathbb{R} \text{ or } \mathbb{T}, \\ u(0,x) = g(x), \end{cases} \tag{3.1}$$

where L is linear symmetric differential operator with constant coefficients and F is the nonlinearity. Note that smooth classical solutions of (3.1) on a time interval $[-\delta, \delta]$ satisfy Duhamel's formula (see Section 3.1 and Exercise 3.1)

$$u(t) = e^{-itL}g - i\int_0^t e^{-i(t-\tau)L}F(u(\tau))d\tau, \quad t \in [-\delta, \delta], \tag{3.2}$$

and that smooth functions satisfying Duhamel's formula also satisfy (3.1).

Definition 3.1 We say that a dispersive PDE (3.1) is locally wellposed in $H^s(K)$, $K = \mathbb{R}$ or $\mathbb{T}$, if there is a Banach space $X_\delta \subset C_t^0 H_x^s([-\delta, \delta] \times K)$ so that for any initial data $g \in H^s(K)$ there exist $\delta = \delta(\|g\|_{H^s}) > 0$ and a unique solution of (3.2) in X_δ. We also require that the data-to-solution map is continuous from H^s to X_δ. We say that the equation is globally wellposed if the local solutions can be extended to a solution in $C_t^0 H_x^s(\mathbb{R} \times K)$.

In the next section, we discuss the local wellposedness of the KdV equation with a smooth potential in high regularity spaces. The addition of a space-time potential in KdV breaks the complete integrability of the equation in the sense that the equation does not satisfy infinitely many conservation laws or that it shares a Lax pair formulation. Our methods are thus purely perturbative. For more details, the reader can consult Hitchin–Segal–Ward [74].

3.1 The energy method

Consider the real-valued KdV equation on $\mathbb{R}$ with a smooth space-time potential V

$$\begin{cases} u_t + u_{xxx} + uu_x + (Vu)_x = 0, & t, x \in \mathbb{R} \\ u(0, x) = g(x) \in H^s(\mathbb{R}), \end{cases} \tag{3.3}$$

with s being a sufficiently large integer. We define

$$M_s = \sup_t \|V\|_{C_x^s}.$$

We say u is a classical solution of (3.3) if

$$u \in C_t^0 H_x^s([-\delta, \delta] \times \mathbb{R}) \cap C_t^1 H_x^{s-3}([-\delta, \delta] \times \mathbb{R})$$

and if u satisfies (3.3) for each x and t. In this section, we prove the existence of smooth classical solutions following Bona–Smith [13] and Kenig [87]. In particular, we prove that if s is large enough, given $g \in H^s$, there exist $\delta = \delta(\|g\|_{H^s})$ and a unique classical solution u on $[-\delta, \delta]$. Moreover, the solution depends continuously on the initial data in H^s topology. Finally, we prove that these solutions can be extended globally-in-time.

It is important to note that the existence of the solutions of (3.3) is not evident in any regularity level because of the derivative nonlinearity. To establish the existence of smooth solutions, we rely on the method of parabolic regularization. More precisely, we first construct the solution of (3.3) with an ϵ-parabolic perturbation. Then we prove that as $\epsilon \to 0$, the limiting solution satisfies (3.3). To achieve that, one needs uniform (in ϵ) bounds for the local existence time and the H^s norm of the solution. This is accomplished by various energy estimates.

3.1.1 A priori bounds

We start with the following energy inequality: if u is a smooth solution to the KdV equation, then there exists $T_0 = T_0(\|g\|_{H^s})$ such that on $[0, T_0]$

$$\|u\|_{H^s} \leq 2\|g\|_{H^s}. \tag{3.4}$$

Indeed

$$\begin{aligned}\partial_t \left\|\partial_x^s u\right\|_{L^2}^2 &= 2\int_{\mathbb{R}} \partial_x^s u_t \partial_x^s u dx \\ &= -2\int_{\mathbb{R}} \partial_x^{s+3} u \partial_x^s u dx - 2\int_{\mathbb{R}} \partial_x^s (u u_x) \partial_x^s u dx - 2\int_{\mathbb{R}} \partial_x^{s+1} (Vu) \partial_x^s u dx. \end{aligned} \tag{3.5}$$

The first term is zero. We estimate the second term. The third term is easier since we can hit V with any number of derivatives up to order $s+1$. We rewrite the second term as follows

$$\begin{aligned}&\int_{\mathbb{R}} \partial_x^s (u u_x) \partial_x^s u dx \\ &= \int_{\mathbb{R}} u \partial_x^{s+1} u \partial_x^s u dx + \int_{\mathbb{R}} \partial_x u \; (\partial_x^s u)^2 dx + \sum_{\substack{k+j=s-1 \\ k,\ j \geq 0}} C_j \int_{\mathbb{R}} \partial_x^s u \left(\partial_x^{k+1} u \partial_x^{j+1} u\right) dx \\ &\qquad = \frac{1}{2} \int_{\mathbb{R}} \partial_x u \; (\partial_x^s u)^2 dx + \sum_{\substack{k+j=s-1 \\ k,\ j \geq 0}} C_j \int_{\mathbb{R}} \partial_x^s u \left(\partial_x^{k+1} u \partial_x^{j+1} u\right) dx,\end{aligned}$$

where to obtain the second equality we used integration by parts. The first term can be estimated as follows

$$\left|\int_{\mathbb{R}} \partial_x u \; (\partial_x^s u)^2 dx\right| \lesssim \|u_x\|_{L^\infty} \|u\|_{H^s}^2.$$

We estimate the second term above using the Cauchy–Schwarz inequality, Corollary 1.8, and the Sobolev embedding

$$\left|\int_{\mathbb{R}} \partial_x^s u \left(\partial_x^{k+1} u \partial_x^{j+1} u\right) dx\right| \lesssim \|u\|_{H^s} \left\|\partial_x^{k+1} u \, \partial_x^{j+1} u\right\|_{L^2} \lesssim \|u_x\|_{L^\infty} \|u\|_{H^s}^2.$$

Thus, for any integer $s \geq 2$, we have

$$\partial_t \|u\|_{H^s}^2 \lesssim \left(\|u_x\|_{L^\infty} + \|V\|_{C_x^{s+1}}\right) \|u\|_{H^s}^2 \lesssim \|u\|_{H^s}^3 + \|V\|_{C_x^{s+1}} \|u\|_{H^s}^2. \tag{3.6}$$

Integrating in time, we obtain

$$\|u(T)\|_{H^s}^2 \leq \|g\|_{H^s}^2 + C \int_0^T \left(\|u(\tau)\|_{H^s}^3 + \|V(\tau)\|_{C_x^{s+1}} \|u(\tau)\|_{H^s}^2\right) d\tau.$$

Let

$$T_0 = \inf \{T : \|u\|_{H^s} \geq 2\|g\|_{H^s}\}.$$

Then on $[0, T_0]$, we have

$$\|u(t)\|_{H^s}^2 \leq \|g\|_{H^s}^2 \left(1 + T_0 \tilde{C} \left(\|g\|_{H^s} + M_{s+1}\right)\right).$$

This implies that we have (3.4) for $t \in [0, T_0]$, where $T_0 \approx \left(\|g\|_{H^s} + M_{s+1}\right)^{-1}$.

Remark 3.2 Note that (3.6) implies via Gronwall's inequality, Exercise 1.6, that

$$\|u\|_{H^s} \leq \|g\|_{H^s} \exp\left(C \int_0^t \left(\|u_x\|_{L^\infty} + M_{s+1}\right) d\tau\right), \tag{3.7}$$

where C is an absolute constant. The advantage of this inequality is that the time that it is valid depends only on the lower index Sobolev norm (H^2 is enough, available by the energy inequality) of the initial data, whereas the time interval in the a priori energy bound depends on the H^s norm. We should also note that (3.7) holds for any real $s \geq 0$, see Exercise 3.3.

3.1.2 Existence and uniqueness

To prove the existence and uniqueness of solutions, we use parabolic regularization. Fix $\epsilon > 0$ and consider the equation

$$\begin{cases} u_t + \epsilon u_{xxxx} + u_{xxx} + uu_x + (Vu)_x = 0, & t, x \in \mathbb{R} \\ u(0, x) = g(x) \in H^s(\mathbb{R}). \end{cases} \tag{3.8}$$

The energy inequalities (3.4) and (3.7) above remain intact since the contribution of the parabolic term in (3.5) is negative. To construct a solution to this equation, we apply Banach's fixed point theorem on the ball

$$X_T = \left\{ u \in C_t^0 H_x^s([0, T] \times \mathbb{R}) : \sup_{t \in [0,T]} \|u(t)\|_{H^s} \leq 2\|g\|_{H^s} \right\},$$

to the operator

$$\Gamma u = e^{-\epsilon t \partial_x^4} g - \int_0^t e^{-\epsilon(t-\tau)\partial_x^4} (u_{xxx} + uu_x + (Vu)_x) dx.$$

It is easy to see that any fixed point that belongs to X_T is a classical solution of the regularized equation (3.8).

We have the following inequalities for the linear group for $t > 0$

$$\begin{aligned}
\left\| e^{-\epsilon t\partial_x^4} u \right\|_{H^s} &\leq \|u\|_{H^s} \\
\left\| e^{-\epsilon t\partial_x^4} \partial_x^3 u \right\|_{H^s} &\lesssim \frac{1}{\epsilon^{3/4} t^{3/4}} \|u\|_{H^s} \\
\left\| e^{-\epsilon t\partial_x^4} (uu_x + (Vu)_x) \right\|_{H^s} &\lesssim \frac{1}{\epsilon^{1/4} t^{1/4}} \left\| u^2 + 2Vu \right\|_{H^s} \\
&\lesssim \frac{1}{\epsilon^{1/4} t^{1/4}} \left(\|u\|_{H^s}^2 + M_s \|u\|_{H^s} \right).
\end{aligned}$$

The first bound follows from the boundedness of the multiplier $e^{-\epsilon t\xi^4}$. The second follows from the inequality

$$\left| \xi^3 e^{-\epsilon t\xi^4} \right| \leq \frac{1}{\epsilon^{3/4} t^{3/4}} \sup_{\eta \in \mathbb{R}} |\eta^3 e^{-\eta^4}| \lesssim \frac{1}{\epsilon^{3/4} t^{3/4}}.$$

The inequality in the third line follows similarly and the last inequality follows from the algebra property of Sobolev spaces. The second bound above is the reason for working with a fourth-order parabolic perturbation. If we had a second-order perturbation, the bound above would not be integrable in t.

Using these inequalities for Γ, we obtain (for $0 \leq t \leq T$ and $u \in X_T$)

$$\begin{aligned}
\|\Gamma u(t)\|_{H^s} &\leq \|g\|_{H^s} + C \int_0^t \frac{\|u(\tau)\|_{H^s}}{\epsilon^{3/4}(t-\tau)^{3/4}} d\tau + C \int_0^t \frac{\|u(\tau)\|_{H^s}^2 + M_s \|u(\tau)\|_{H^s}}{\epsilon^{1/4}(t-\tau)^{1/4}} d\tau \\
&\leq \|g\|_{H^s} + C\epsilon^{-3/4} t^{1/4} \|g\|_{H^s} + C\epsilon^{-1/4} t^{3/4} \left(\|g\|_{H^s}^2 + M_s \|g\|_{H^s} \right) \\
&\leq 2\|g\|_{H^s},
\end{aligned}$$

if $T \leq T_1(\epsilon, \|g\|_{H^s})$. Estimating the differences $\Gamma(u) - \Gamma(v)$ in a similar manner implies that Γ is a construction in X_T. Also note that since the multiplier $e^{-\epsilon t\xi^4}$ maps L^2 to $H^\infty =: \cap_{s \geq 0} H^s$, these local solutions are in H^∞ for any $t \in (0, T]$.

Although the existence time depends on ϵ, by iterating these local solutions using the energy inequality (3.4) we obtain a solution, u^ϵ, valid in the time interval $[0, T_0]$, $T_0 = T_0(\|g\|_{H^s})$. Moreover, using the equation, we have $u^\epsilon \in C_t^1 H_x^{s-4}([0, T_0] \times \mathbb{R})$. From now on, we will denote T_0 by δ.

Now we need to prove that u^ϵ converges to a solution of the KdV equation as ϵ tends to zero. To do this, we first show that u^ϵ is Cauchy in $C_t^0 L_x^2([0, \delta] \times \mathbb{R})$. Take $0 < \epsilon < \epsilon'$ and consider the corresponding solutions. Using the equation for ϵ and ϵ', we have

$$\begin{aligned}\partial_t \left\|u^\epsilon - u^{\epsilon'}\right\|_{L^2}^2 &= -2\epsilon' \int_{\mathbb{R}} \left(u^\epsilon - u^{\epsilon'}\right) \partial_x^4 \left(u^\epsilon - u^{\epsilon'}\right) - 2(\epsilon - \epsilon') \int_{\mathbb{R}} \left(u^\epsilon - u^{\epsilon'}\right) \partial_x^4 u^\epsilon \\ &\quad - \int_{\mathbb{R}} \left(u^\epsilon - u^{\epsilon'}\right) \partial_x^3 \left(u^\epsilon - u^{\epsilon'}\right) \\ &\quad - \int_{\mathbb{R}} \left(u^\epsilon - u^{\epsilon'}\right) \partial_x \left[\left(u^\epsilon - u^{\epsilon'}\right)\left(u^\epsilon + u^{\epsilon'} + 2V\right)\right] \\ &= -2\epsilon' \int_{\mathbb{R}} \left(u^\epsilon - u^{\epsilon'}\right) \partial_x^4 \left(u^\epsilon - u^{\epsilon'}\right) - 2(\epsilon - \epsilon') \int_{\mathbb{R}} \left(u^\epsilon - u^{\epsilon'}\right) \partial_x^4 u^\epsilon \\ &\quad - \frac{1}{2} \int_{\mathbb{R}} \left(u^\epsilon - u^{\epsilon'}\right)^2 \partial_x \left(u^\epsilon + u^{\epsilon'} + 2V\right) \\ &\leq -2(\epsilon - \epsilon') \int_{\mathbb{R}} \left(u^\epsilon - u^{\epsilon'}\right) \partial_x^4 u^\epsilon \\ &\quad - \frac{1}{2} \int_{\mathbb{R}} \left(u^\epsilon - u^{\epsilon'}\right)^2 \partial_x \left(u^\epsilon + u^{\epsilon'} + 2V\right).\end{aligned}$$

The second equality follows by noting that the third integral is zero. The last inequality follows by the inequality

$$-2\epsilon' \int_{\mathbb{R}} \left(u^\epsilon - u^{\epsilon'}\right) \partial_x^4 \left(u^\epsilon - u^{\epsilon'}\right) = -2\epsilon' \int_{\mathbb{R}} \left[\partial_x^2 \left(u^\epsilon - u^{\epsilon'}\right)\right]^2 \leq 0.$$

We estimate the remaining terms using the Cauchy–Schwarz inequality and Sobolev embedding to obtain

$$\begin{aligned}&\partial_t \left\|u^\epsilon - u^{\epsilon'}\right\|_{L^2}^2 \lesssim \\ &\qquad |\epsilon - \epsilon'| \left\|u^\epsilon - u^{\epsilon'}\right\|_{L^2} \|u^\epsilon\|_{H^4} + \left\|u^\epsilon - u^{\epsilon'}\right\|_{L^2}^2 \left(\|u^\epsilon\|_{H^2} + \left\|u^{\epsilon'}\right\|_{H^2} + M_1\right).\end{aligned}$$

This implies, by cancelling the L^2 norm of the difference, that

$$\partial_t \left\|u^\epsilon - u^{\epsilon'}\right\|_{L^2} \lesssim |\epsilon - \epsilon'| \, \|u^\epsilon\|_{H^4} + \left\|u^\epsilon - u^{\epsilon'}\right\|_{L^2} \left(\|u^\epsilon\|_{H^2} + \left\|u^{\epsilon'}\right\|_{H^2} + M_1\right).$$

Using the a priori bound

$$\left\|u^{\epsilon'}\right\|_{H^4} \lesssim \|g\|_{H^4}$$

and Gronwall's inequality, we obtain

$$\sup_{[0,\delta]} \left\|u^\epsilon - u^{\epsilon'}\right\|_{L^2} \lesssim |\epsilon - \epsilon'|.$$

Therefore, u^ϵ is Cauchy in $C_t^0 L_x^2([0,\delta] \times \mathbb{R})$. By interpolation with $L^\infty H^s$, u^ϵ is also Cauchy in $C_t^0 H_x^r([0,\delta] \times \mathbb{R})$ for any $r \in [0, s)$. Moreover, using the equation we conclude that $\partial_t u^\epsilon$ is Cauchy in $C_t^0 H_x^{r-4}([0,\delta] \times \mathbb{R})$. Therefore, the limiting function $u \in C_t^0 H_x^r([0,\delta] \times \mathbb{R}) \cap C_t^1 H_x^{r-4}([0,\delta] \times \mathbb{R})$ solves (3.3) (by taking pointwise limits). Also note using Fatou's lemma that, since u^ϵ is bounded in H^s and converges to u in L^2, $u \in L_t^\infty H_x^s([0,\delta]\times\mathbb{R})$.

We now continue with uniqueness. Consider (3.3) with initial data g_1 and g_2 in H^s. Let u and v be the corresponding solutions valid in a common time interval $[0,\delta]$. By using the equation as above and Sobolev embedding, we obtain

$$\partial_t\|u-v\|_{L^2}^2 = -2\int_{\mathbb{R}}(u-v)\partial_x^3(u-v) - \int_{\mathbb{R}}(u-v)\partial_x[(u-v)(u+v+2V)]$$
$$\lesssim \|u-v\|_{L^2}^2\left(\|u\|_{H^2}+\|v\|_{H^2}+M_1\right).$$

Therefore, by Gronwall's inequality we obtain

$$\|u-v\|_{L^2} \lesssim \|g_1-g_2\|_{L^2} \tag{3.9}$$

on $[0,\delta]$. This implies uniqueness.

It remains to prove the continuous dependence on initial data, and that $u \in C_t^0H_x^s([0,\delta]\times\mathbb{R})$. This also implies that $u \in C_t^1H_x^{s-3}([0,\delta]\times\mathbb{R})$ by the equation. To do this, we regularize the initial data as follows

$$g^\eta := g * \varphi_\eta.$$

Here φ is a Schwartz function with mean 1 satisfying $\partial_\xi^k\widehat{\varphi}(0) = 0$ for all $k > 0$ (take $\widehat{\varphi}$ constant 1 in a neighborhood of the origin), and $\varphi_\eta(x) = \frac{1}{\eta}\varphi(\frac{x}{\eta})$. Since φ_η is an approximate identity, g^η converges to g in H^s as $\eta\to 0$. Further, note that $\|g^\eta\|_{H^s} \lesssim \|g\|_{H^s}$, where the implicit constant is independent of η. Let u^η be the solution with initial data g^η on $[0,\delta]$ constructed above.

The following lemma contains all technical estimates we need to finish the proof:

Lemma 3.3 *Consider the solutions u^η and v^η constructed above on $[0,\delta]$ starting with data g, h respectively. We claim that:*
(i) $\|u^\eta\|_{H^{s+1}} \lesssim 1/\eta$,
(ii) u^η converges to u in $L_t^\infty H_x^s([0,\delta]\times\mathbb{R})$,
(iii) $\|u^\eta - v^\eta\|_{H^s} \lesssim e^{C\delta/\eta}\|g-h\|_{H^s}$,
(iv) Assume that g_n converges to g in H^s. Let u_n and u_n^η be the solutions corresponding to the initial data g_n and g_n^η respectively. Then

$$\sup_n \left\|u_n^\eta - u_n\right\|_{H^s} \to 0 \quad \text{as } \eta\to 0.$$

The bound (i) and interpolation imply that

$$u^\eta \in C_t^0L_x^2([0,\delta]\times\mathbb{R}) \cap L_t^\infty H_x^{s+1}([0,\delta]\times\mathbb{R}) \subset C_t^0H_x^s([0,\delta]\times\mathbb{R}).$$

This implies that $u \in C_t^0 H_x^s([0,\delta]\times\mathbb{R})$, since by (ii) u^η converges to u uniformly on $[0,\delta]$.

The lemma also implies continuous dependence on initial data as follows. Assume that g_n converges to g in H^s. Construct the regularized solutions as in the lemma. Using triangle inequality and (iii), we have (for $t \in [0,\delta]$)

$$\begin{aligned}\|u - u_n\|_{H^s} &\le \|u - u^\eta\|_{H^s} + \left\|u^\eta - u_n^\eta\right\|_{H^s} + \left\|u_n^\eta - u_n\right\|_{H^s}\\ &\lesssim \|u - u^\eta\|_{H^s} + e^{C\delta/\eta}\,\|g - g_n\|_{H^s} + \sup_j \left\|u_j^\eta - u_j\right\|_{H^s}.\end{aligned}$$

Given $\epsilon > 0$, fix η sufficiently small so that in light of (ii) and (iv) we have

$$\|u - u_n\|_{H^s} \lesssim \epsilon + e^{C\delta/\eta}\|g - g_n\|_{H^s} + \epsilon.$$

Taking $n \to \infty$ finishes the proof. It remains to prove the lemma.

Proof of Lemma 3.3 (i) Note that by (3.7) we have on $[0,\delta]$

$$\begin{aligned}\|u^\eta\|_{H^{s+1}} &\lesssim \|g^\eta\|_{H^{s+1}} \exp\left(C \int_0^t \left(\|u_x^\eta\|_\infty + M_{s+1}\right) d\tau\right)\\ &\lesssim \|g^\eta\|_{H^{s+1}} \exp\left(C\delta\left(\|g^\eta\|_{H^s} + M_{s+1}\right)\right).\end{aligned}$$

Therefore, it suffices to prove (i) at time 0. Indeed

$$\|g^\eta\|_{H^{s+1}} = \|\langle\xi\rangle\widehat{\varphi}(\eta\xi)\langle\xi\rangle^s\widehat{g}(\xi)\|_{L^2} \lesssim \|\langle\xi\rangle\widehat{\varphi}(\eta\xi)\|_{L^\infty}\|g\|_{H^s} \lesssim 1/\eta.$$

(ii) We first prove that for $0 < \eta' < \eta$, we have

$$\|u^\eta - u^{\eta'}\|_{L^2} = o(\eta^s). \tag{3.10}$$

By (3.9), it suffices to prove this at time zero. We have

$$\|g^\eta - g^{\eta'}\|_{L^2}^2 = \int_{\mathbb{R}} \left|\widehat{\varphi}(\eta\xi) - \widehat{\varphi}(\eta'\xi)\right|^2 \langle\xi\rangle^{-2s}|\widehat{g}(\xi)|^2\langle\xi\rangle^{2s}d\xi.$$

By Taylor expansion and the fact that the derivatives of $\widehat{\varphi}$ vanishes at zero, we have

$$\widehat{\varphi}(\eta\xi) = 1 + O\left(\eta^s\xi^s \sup_{[0,\eta\xi]} |\partial^s\widehat{\varphi}|\right).$$

Thus, we have

$$\left\|g^\eta - g^{\eta'}\right\|_{L^2}^2 \lesssim \eta^{2s} \int_{\mathbb{R}} \left\|\partial^s\widehat{\varphi}\right\|_{L^\infty([0,\eta\xi])}^2 |\widehat{g}(\xi)|^2\langle\xi\rangle^{2s}d\xi. \tag{3.11}$$

Since

$$\partial^s\widehat{\varphi}(0) = 0,$$

the inequality (3.10) follows from the dominated convergence theorem.

Interpolating (3.10) with the bound (i), we obtain $\|u^\eta - u^{\eta'}\|_{H^s} = o(1)$ as η, η' go to zero. This implies that u^η is a convergent sequence in $L_t^\infty H_x^s([0,\delta]\times\mathbb{R})$. By (3.9)

$$\|u^\eta - u\|_{L^2} \lesssim \|g^\eta - g\|_{L^2} \to 0$$

as $\eta \to 0$; therefore, u is the limit of u^η also in H^s.
(iii) Using the equation, we estimate

$$\begin{aligned}
\partial_t \|\partial_x^s(u^\eta - v^\eta)\|_{L^2}^2 &= 2\int_{\mathbb{R}} \partial_x^s(u^\eta - v^\eta)_t \partial_x^s(u^\eta - v^\eta)dx \\
&= -2\int_{\mathbb{R}} \partial_x^{s+3}(u^\eta - v^\eta)\partial_x^s(u^\eta - v^\eta)dx \\
&\quad - \int_{\mathbb{R}} \partial_x^{s+1}\left[(u^\eta - v^\eta)(u^\eta + v^\eta + 2V)\right]\partial_x^s(u^\eta - v^\eta)dx \\
&\lesssim \|u^\eta - v^\eta\|_{H^s}^2 \left(\|u^\eta\|_{H^{s+1}} + \|v^\eta\|_{H^{s+1}} + M_{s+1}\right) \\
&\lesssim \frac{1}{\eta}\|u^\eta - v^\eta\|_{H^s}^2.
\end{aligned}$$

The first inequality follows since the first summand is zero and the second one can be estimated by considering the cases when $s+1$ derivatives hit $u^\eta + v^\eta$ and $u^\eta - v^\eta$. The second inequality follows from (i). This implies (iii) by Gronwall's inequality.
(iv) Since

$$\left\|u_n^\eta - u_n\right\|_{H^s} = \lim_{\eta'\to 0}\left\|u_n^\eta - u_n^{\eta'}\right\|_{H^s},$$

it suffices to prove that

$$\sup_{\eta'<\eta,\, n}\left\|u_n^\eta - u_n^{\eta'}\right\|_{L^2} = o(\eta^s).$$

Interpolation with the bound (i) yields the claim.

Using (3.9) and (3.11), it is enough to show that

$$\sup_n \int_{\mathbb{R}} \|\partial^s\varphi\|_{L^\infty([0,\eta\xi])}^2 \left|\widehat{g_n}(\xi)\right|^2 \langle\xi\rangle^{2s} d\xi = o(1).$$

Indeed

$$\begin{aligned}
\sup_n \int_{\mathbb{R}} &\|\partial^s\widehat{\varphi}\|_{L^\infty([0,\eta\xi])}^2 \left|\widehat{g_n}\right|^2 \langle\xi\rangle^{2s} d\xi \\
&\le \int_{\mathbb{R}} \|\partial^s\widehat{\varphi}\|_{L^\infty([0,\eta\xi])}^2 \left|\widehat{g}\right|^2 \langle\xi\rangle^{2s} d\xi \\
&\qquad + \sup_n \int_{\mathbb{R}} \|\partial^s\widehat{\varphi}\|_{L^\infty([0,\eta\xi])}^2 \left|\widehat{g_n} - \widehat{g}\right|^2 \langle\xi\rangle^{2s} d\xi.
\end{aligned}$$

The first integral goes to zero by the Lebesgue dominated convergence theorem. For the second, given $\epsilon > 0$, choose N so that $\|g_n - g\|_{H^s} < \epsilon$ for all $n > N$, and estimate the first N terms by the dominated convergence theorem. □

Remark 3.4 In the proof above, we assumed that s is large enough. In the original paper of Bona–Smith [13], wellposedness was obtained in H^s, $s \geq 2$. The local wellposedness range was extended to H^s, $s > \frac{3}{2}$, by Saut–Temam [126] and Kato [82]. Note that the range of s is dictated by the Sobolev embedding theorem since $\|u_x\|_{L^\infty} \lesssim \|u\|_{H^s}$, $s > \frac{3}{2}$. To obtain wellposedness results beyond this range, one needs to use the smoothing effects of time averaging either by employing space-time norms as in Strichartz estimates or by normal form transformations. We will discuss these issues in the next three sections.

Remark 3.5 The construction above is valid also in the periodic case; see Bona–Smith [13]. The only required modification is in the definition of the regularized function g^η. Instead of taking a convolution, we define g^η on the Fourier side as

$$\widehat{g^\eta}(k) = \widehat{g}(k)\psi(\eta k),$$

where ψ is a function on $\mathbb{R}$ satisfying the same properties as $\widehat{\varphi}$.

It remains to prove that the solutions constructed above can be defined globally-in-time. It is a well-known fact in the literature that smooth solutions of the KdV equation (without potential) satisfy infinitely many conservation laws; see Hitchin–Segal–Ward [74]. A sample includes the following

$$I_1(t) = \int_{\mathbb{R}} u(t,x)dx = I_1(0),$$

$$I_2(t) = \int_{\mathbb{R}} u^2(t,x)dx = I_2(0),$$

$$I_3(t) = \int_{\mathbb{R}} \left(u_x^2 - \frac{1}{3}u^3\right) dx = I_3(0),$$

$$I_4(t) = \int_{\mathbb{R}} \left(u_{xx}^2 - \frac{5}{3}uu_x^2 - \frac{5}{12}u^4\right) dx = I_4(0).$$

This can be verified directly by taking the time derivative of the above quantities and showing that $\partial_t I_j = 0$, $j = 1,2,3,4$. Each conservation law along with Sobolev embedding provides an a priori bound

$$\|u(t)\|_{H^s} \le C_s \|g\|_{H^s} \tag{3.12}$$

for integer $s \ge 0$. In general for a given H^s solution, to make sense of the time differentiation and then integration by parts, we need the solutions to live in a smoother than H^s space (H^{s+3} suffices for the KdV equation). One resolves this minor problem by considering smooth solutions as follows. Let $g \in H^s$ and as before construct g^η smooth such that $g^\eta \to g$ in H^s. By continuous dependence, we know that $u^\eta \to u$ in H^s. Since u^η is smooth, it satisfies the a priori bound (3.12). But then

$$\|u\|_{H^s} \le \|u^\eta - u\|_{H^s} + \|u^\eta\|_{H^s} \lesssim \|u^\eta - u\|_{H^s} + \|g^\eta\|_{H^s} \lesssim \|g\|_{H^s}$$

by taking $\eta \to 0$. Thus, u satisfies the a priori bound. But then we can iterate the local solution and reach any time interval $[0,T]$. To see this, use (3.12), and choose $\delta = \delta(C_s\|g\|_{H^s})$. We have a local solution on $[0,\delta]$ satisfying (3.12) on the same interval. We can now solve the KdV equation with initial data $u(\delta)$ on $[\delta, 2\delta]$ because of the choice of δ. By continuity, we can glue these solutions together and thus u solves the KdV equation on $[0,2\delta]$, and on this interval it satisfies (3.12). We continue with this uniform time step to cover $[0,T]$.

Remark 3.6 Notice that (in the absence of a conserved quantity) one cannot iterate the a priori bound coming from the energy inequality (3.4). This is because, as $\|u\|_{H^s}$ grows, going from one time interval to another, the time intervals shrinks. Thus, it is possible that the sequence of times shrinks in such a way that it approaches a finite time limit and the process stops.

3.1.3 Growth bounds for KdV with potential

In the case of the KdV equation with a potential, the quantities I_j are not conserved anymore. However, they still allow us to get exponential bounds for H^s norms for all $s \ge 0$ integer. Indeed

$$\frac{d}{dt} I_2(t) = 2\int_{\mathbb{R}} u u_t \, dx = \int_{\mathbb{R}} u^2 V_x \, dx \le M_1 I_2(t).$$

This implies that

$$\|u(t)\|_{L^2} \le \|g\|_{L_2} e^{tM_1/2}. \tag{3.13}$$

The proof for higher order norms is similar but more complicated. We give the details for I_3. Note that by Gagliardo–Nirenberg inequality and (3.13), we have

$$I_3 \ge \|u_x\|_{L^2}^2 - C\|u_x\|_{L^2}^{1/2}\|u\|_{L^2}^{5/2} \ge \|u_x\|_{L^2}^2 - C\|u_x\|_{L^2}^{1/2} e^{t5M_1/4}.$$

This implies that

$$\|u_x\|_{L^2} \lesssim \sqrt{I_3} + e^{Ct}. \tag{3.14}$$

On the other hand, using the equation (note that the terms without the potential cancel out since I_3 is conserved for the KdV equation), (3.13), (3.14), and the Gagliardo–Nirenberg inequality, we obtain

$$\begin{aligned}\frac{d}{dt}I_3(t) &= -2\int_{\mathbb{R}} V_x u_x^2\,dx - 2\int_{\mathbb{R}} V_{xx} u u_x\,dx + \frac{2}{3}\int_{\mathbb{R}} u^3 V_x\,dx \\ &\lesssim \|u_x\|_{L^2}^2 + \|u_x\|_{L^2}\|u\|_{L^2} + \|u_x\|_{L^2}^{1/2}\|u\|_{L^2}^{5/2} \\ &\lesssim I_3(t) + I_3(t)^{1/4}e^{Ct} + e^{Ct} \lesssim I_3(t) + e^{Ct}.\end{aligned}$$

This implies that I_3 grows at most exponentially, and hence

$$\|u_x\|_{L^2} \lesssim e^{Ct}.$$

A similar calculation using I_4 and the bounds above gives an exponential bound for $\|u_{xx}\|_{L^2}$. Using this in (3.7), one obtains exponential bounds for all higher order Sobolev norms H^s, $s \geq 2$ real.

3.2 Oscillatory integral method

In this section, we establish the local wellposedness of the KdV equation on the real line

$$\begin{cases} u_t + u_{xxx} + uu_x = 0, & t, x \in \mathbb{R} \\ u(0,x) = g(x) \in H^s(\mathbb{R}), \end{cases} \tag{3.15}$$

for $\frac{3}{4} < s < 1$ (the same method works for all $s > \frac{3}{4}$). The proof relies on the dispersive decay, smoothing, and maximal function estimates we obtained for the Airy equation on $\mathbb{R}$ in Chapter 2. As such this method is not suitable for nonlinear dispersive PDEs on bounded domains. We base our approach on a series of papers by Kenig–Ponce–Vega; see, e.g., [88, 89], and the references therein.

To motivate the choice for the space X we work with, we start with the following:

Lemma 3.7 *For $s \in (\frac{1}{2}, 1)$ and $0 < \delta \lesssim 1$, we have*

$$\|uu_x\|_{L^2_{t\in[0,\delta]}H^s_x} \lesssim \delta^{1/4}\|u_x\|_{L^4_t L^\infty_x}\|u\|_{L^\infty_t H^s_x} + \|u\|_{L^2_x L^\infty_t}\|D^s u_x\|_{L^\infty_x L^2_t} + \delta^{\frac{1}{2}}\|u\|^2_{L^\infty_t H^s_x}.$$

Proof First note that

$$\|uu_x\|_{H^s_x} = \|J^s(uu_x)\|_{L^2_x} \le \|J^s(uu_x) - uJ^s u_x\|_{L^2_x} + \|uJ^s u_x\|_{L^2_x} =: I + II.$$

Using Lemma 1.9, we obtain

$$I \lesssim \|u_x\|_{L^\infty_x} \|u\|_{H^s}.$$

Note that

$$\begin{aligned} II &\le \|uD^s u_x\|_{L^2_x} + \|u(J^s - D^s)u_x\|_{L^2_x} \\ &\lesssim \|uD^s u_x\|_{L^2_x} + \|u\|_{L^\infty_x}\|u\|_{L^2_x} \lesssim \|uD^s u_x\|_{L^2_x} + \|u\|^2_{H^s}. \end{aligned}$$

The second inequality follows from the boundedness of the multiplier (recall that $s \in [0, 1]$)

$$m(\xi) = \xi \left[\left(1 + |\xi|^2\right)^{\frac{s}{2}} - |\xi|^s \right],$$

and the third by Sobolev embedding. Combining these bounds and then using Hölder's inequality yield the statement

$$\begin{aligned} \|uu_x\|_{L^2_{t\in[0,\delta]}H^s_x} &\lesssim \left\| \|u_x\|_{L^\infty_x}\|u\|_{H^s} \right\|_{L^2_{t\in[0,\delta]}} + \left\| \|u\|^2_{H^s} \right\|_{L^2_{t\in[0,\delta]}} + \|uD^s u_x\|_{L^2_x L^2_{t\in[0,\delta]}} \\ &\lesssim \delta^{1/4}\|u_x\|_{L^4_t L^\infty_x}\|u\|_{L^\infty_t H^s_x} + \delta^{\frac{1}{2}}\|u\|^2_{L^\infty_t H^s_x} + \|u\|_{L^2_x L^\infty_t}\|D^s u_x\|_{L^\infty_x L^2_t}. \end{aligned}$$

Note that we changed the order of the norms in the last term of the first line. □

Now we are ready to prove that the KdV equation is locally wellposed on $H^s(\mathbb{R})$ for $s > 3/4$. To do that, we apply Banach's fixed point theorem on the ball (for sufficiently small $\delta = \delta(\|g\|_{H^s})$)

$$B_\delta = \left\{ u \in X \cap C^0_t H^s_x([0, \delta] \times \mathbb{R}) : \|u\|_X \le M\|g\|_{H^s} \right\},$$

to the operator

$$\Gamma u(t) = W_t g(x) - \int_0^t W_{t-\tau}(u(\tau)u_x(\tau))d\tau,$$

where

$$\|u\|_X = \max\left(\|u\|_{L^\infty_{t\in[0,\delta]}H^s_x}, \|u_x\|_{L^4_{t\in[0,\delta]}L^\infty_x}, \|u\|_{L^2_x L^\infty_{t\in[0,\delta]}}, \|D^s u_x\|_{L^\infty_x L^2_{t\in[0,\delta]}} \right).$$

Recall that $W_t = e^{-t\partial^3_x}$, W_t is unitary in H^s spaces, and that W_t commutes with J^s, D^s, and the Hilbert transform H. Note that the norms appearing in the definition of X are the ones on the right-hand side of the inequality in Lemma 3.7.

We only prove that Γ maps B_δ into itself provided that $\delta = \delta(\|g\|_{H^s})$ is sufficiently small and M is sufficiently large. Estimates on the differences can be obtained similarly. We start with $\|\Gamma u\|_{L^\infty_{t\in[0,\delta]} H^s_x}$

$$
\begin{aligned}
\|\Gamma u\|_{L^\infty_t H^s_x} &\lesssim \|g\|_{H^s} + \left\| \int_0^t W_{t-\tau}(uu_x)d\tau \right\|_{L^\infty_t H^s_x} \\
&\leq \|g\|_{H^s} + \left\| \int_0^t \|uu_x\|_{H^s_x} d\tau \right\|_{L^\infty_t} \\
&\lesssim \|g\|_{H^s} + \delta^{1/2}\|uu_x\|_{L^2_t H^s_x} \lesssim \|g\|_{H^s} + \delta^{1/2}\|u\|^2_X .
\end{aligned}
$$

We used Lemma 3.7 in the last inequality.

Similarly, using $\partial_x = DH$, where H is the Hilbert transform, and the fact that W_t commutes with D and H, we obtain

$$
\begin{aligned}
\|\partial_x \Gamma u\|_{L^4_{t\in[0,\delta]} L^\infty_x} &\lesssim \|\partial_x W_t g\|_{L^4_t L^\infty_x} + \left\| \partial_x \int_0^t W_{t-\tau}(uu_x)d\tau \right\|_{L^4_t L^\infty_x} \\
&\leq \left\| D^{1/4} W_t D^{3/4} H g \right\|_{L^4_t L^\infty_x} + \left\| \int_0^t \left\| D^{1/4} W_t W_{-\tau} D^{3/4} H(uu_x) \right\|_{L^\infty_x} d\tau \right\|_{L^4_t} \\
&\leq \left\| D^{3/4} H g \right\|_{L^2_x} + \int_0^\delta \left\| D^{1/4} W_t W_{-\tau} D^{3/4} H(uu_x) \right\|_{L^4_t L^\infty_x} d\tau \\
&\leq \left\| D^{3/4} H g \right\|_{L^2_x} + \int_0^\delta \left\| D^{3/4} H(uu_x) \right\|_{L^2_x} d\tau \\
&\lesssim \|g\|_{H^s} + \delta^{1/2}\|uu_x\|_{L^2_t H^s_x} \lesssim \|g\|_{H^s} + \delta^{1/2}\|u\|^2_X .
\end{aligned}
$$

In the third and forth inequalities, we used Strichartz inequality (2.8), and in the last inequality, we used Lemma 3.7.

By the maximal function inequality (Theorem 2.6), we have

$$
\begin{aligned}
\|\Gamma u\|_{L^2_x L^\infty_{t\in[0,\delta]}} &\leq \|W_t g\|_{L^2_x L^\infty_t} + \left\| \int_0^t |W_{t-\tau}(uu_x)|\, d\tau \right\|_{L^2_x L^\infty_t} \\
&\lesssim \|g\|_{H^s} + \int_0^\delta \|W_t W_{-\tau}(uu_x)\|_{L^2_x L^\infty_t} d\tau \\
&\lesssim \|g\|_{H^s} + \int_0^\delta \|uu_x\|_{H^s_x} d\tau \\
&\lesssim \|g\|_{H^s} + \delta^{1/2}\, \|uu_x\|_{L^2_t H^s_x} \lesssim \|g\|_{H^s} + \delta^{1/2}\|u\|^2_X .
\end{aligned}
$$

Finally, we estimate $\|D^s \partial_x \Gamma u\|_{L^\infty_x L^2_{t\in[0,\delta]}}$ using the Kato smoothing estimate

(Theorem 2.5)

$$\begin{aligned}
\|D^s\partial_x\Gamma u\|_{L^\infty_x L^2_{t\in[0,\delta]}} &\le \|\partial_x W_t D^s g\|_{L^\infty_x L^2_t} + \left\| \int_0^t |D^s\partial_x W_{t-\tau}(uu_x)|\, d\tau \right\|_{L^\infty_x L^2_t} \\
&\lesssim \|g\|_{H^s} + \int_0^\delta \|\partial_x W_t W_{-\tau} D^s(uu_x)\|_{L^\infty_x L^2_t}\, d\tau \\
&\lesssim \|g\|_{H^s} + \int_0^\delta \|uu_x\|_{H^s_x}\, d\tau \\
&\lesssim \|g\|_{H^s} + \delta^{1/2}\|uu_x\|_{L^2_t H^s_x} \lesssim \|g\|_{H^s} + \delta^{1/2}\|u\|_X^2.
\end{aligned}$$

We thus obtain the following estimate

$$\|\Gamma u\|_X \lesssim \|g\|_{H^s} + \delta^{1/2}\|u\|_X^2.$$

Therefore, one can close the argument by choosing M large and δ small, provided that $\Gamma u \in C^0_t H^s_x([0,\delta] \times \mathbb{R})$. We only prove the continuity at time 0. The proof is similar for each time because of the group structure of W_t. By the bounds above and the continuity of W_t in H^s, we have

$$\|\Gamma u(t) - g\|_{H^s} \lesssim \|W_t g - g\|_{H^s} + t^{1/2}\|u\|_X^2 \to 0, \quad \text{as } t \to 0,$$

which finishes the proof.

Since we have proved the local wellposedness by a contraction argument, the method also implies continuous (Lipschitz) dependence on the initial data.

Remark 3.8 Suppose that one proves existence and uniqueness in B_δ which is a fixed ball in the space X. One can then easily extend the uniqueness to the whole space X by shrinking time by a fixed amount. Indeed, by shrinking time to δ' we get existence and uniqueness in a larger ball $B_{\delta'}$. Now assume that there are two different solutions, one staying in the ball B_δ and one separating after hitting the boundary at some time $|t| < \delta'$. This is already a contradiction by the uniqueness in $B_{\delta'}$.

Finally, the H^1 a priori bound coming from the conservation laws extends this local solution globally-in-time in H^1 as described in the previous section.

We note that the oscillatory integral method is very efficient on unbounded domains where the dispersion is in full effect, and it has been applied to various dispersive models with derivative nonlinearities, such as the generalized KdV equations, derivative NLS equations, and Zakharov system; see, e.g., Kenig–Ponce–Vega [88, 89, 90].

As mentioned above, one drawback of this method is the fact that it relies on the dispersive estimates that are not true over compact domains, in particular over $\mathbb{T}$. In the next section, we demonstrate a method that can be used to study local wellposedness on both $\mathbb{R}$ and $\mathbb{T}$.

3.3 Restricted norm method

In Section 3.2, using oscillatory integral methods, we proved the local wellposedness of the KdV equation (3.15) in $H^s(\mathbb{R})$, $s > \frac{3}{4}$. In the early nineties, Bourgain [15, 16] introduced a new class of spaces that take into account the dispersive symbol of the equation. Note that the space-time Fourier transform of the solution of the Airy equation is a distribution supported on the curve $\tau = \xi^3$. The idea is based on the observation that for short times the space-time Fourier transform of the KdV solution will essentially be supported in a small neighborhood of this curve. To this end, we introduce the norm

$$\|u\|_{X^{s,b}} = \left\|\langle\xi\rangle^s\langle\tau - \xi^3\rangle^b\widehat{u}(\tau,\xi)\right\|_{L^2_{\xi,\tau}}.$$

Variations of this norm for the wave equation have appeared before in Beals [4] and Klainerman–Machedon [96].

Using these spaces, Bourgain extended the local wellposedness theory of the KdV equation to the L^2 level both on $\mathbb{R}$ and $\mathbb{T}$. Because of the L^2 norm conservation, these solutions are globally defined. Later, almost sharp results were obtained by Kenig–Ponce–Vega in [91]. In particular, they obtained local wellposedness in $H^s(\mathbb{R})$ for $s > -\frac{3}{4}$ and in $H^s(\mathbb{T})$ for $s > -\frac{1}{2}$. The local and global wellposedness in $H^{-\frac{1}{2}}(\mathbb{T})$ was obtained by Colliander–Keel–Staffilani–Takaoka–Tao in [43]. They also obtained global wellposedness in $H^s(\mathbb{R})$ for $s > -\frac{3}{4}$. The endpoint local wellposedness result for $\mathbb{R}$ was obtained by Christ–Colliander–Tao in [38]. The global wellposedness at $H^{-\frac{3}{4}}(\mathbb{R})$ was obtained by Guo [72]. Finally, in [81], Kappeler and Topalov extended the local and global wellposedness result on the torus to $H^{-1}(\mathbb{T})$ using the integrability properties of the equation.

3.3.1 L^2 solutions of KdV on the real line

In this section, we construct L^2 solutions of the KdV equation (3.15) on $\mathbb{R}$. By the conservation law, these solutions are global-in-time. Let $X^{s,b}$ be the Banach space of functions on $\mathbb{R}\times\mathbb{R}$ defined by the norm

$$\|u\|_{X^{s,b}} = \left\|\langle\xi\rangle^s\langle\tau - \xi^3\rangle^b\widehat{u}(\tau,\xi)\right\|_{L^2_{\xi,\tau}} = \left\|W_{-t}u\right\|_{H^s_xH^b_t}. \tag{3.16}$$

Here the $\|f\|_{H^s_x H^b_t}$ norm is not the iterated norm in the usual sense. Sometimes, it is useful to rewrite it as follows

$$\|f\|_{H^s_x H^b_t} = \left\|\widehat{f}(\tau,\xi)\langle\tau\rangle^b\langle\xi\rangle^s\right\|_{L^2_{\xi,\tau}} = \left\|\left\|f\left(t,\widehat{\xi}\right)\langle\xi\rangle^s\right\|_{H^b_t}\right\|_{L^2_\xi}. \tag{3.17}$$

It is easy to see that $X^{-s,-b}$ is the dual space of $X^{s,b}$. Since the contraction argument will be in a time interval $[-\delta,\delta]$ with $\delta \leq 1$, we also define the restricted $X^{s,b}$ space, $X^{s,b}_\delta$, as the equivalent classes of functions that agree on $[-\delta,\delta]$ with the norm

$$\|u\|_{X^{s,b}_\delta} = \inf_{\tilde{u}=u,\, t\in[-\delta,\delta]} \|\tilde{u}\|_{X^{s,b}}.$$

The contraction argument is applied to the operator

$$\Gamma u = \eta(t)W_t g - \eta(t)\int_0^t W_{t-s}(uu_x)ds,$$

on the ball

$$B_\delta = \left\{u \in X^{0,b}_\delta : \|u\|_{X^{0,b}_\delta} \leq C\|g\|_{L^2}\right\},$$

where η is a C^∞_0 function satisfying $\eta(t) = 1, t \in [-1,1]$. Since $\delta \leq 1$, a fixed point of Γ gives us a solution of the KdV equation on $[-\delta,\delta]$. The continuity of these solutions follows from:

Lemma 3.9 *For any $b > \frac{1}{2}$, $X^{s,b}_\delta$ embeds into $C^0_t H^s_x([-\delta,\delta]\times\mathbb{R})$.*

Proof Let $v(t,x) = u(t,x)$ for all $|t| \leq \delta$. By Fourier inversion and then a change of variable, we have (for $|t| \leq \delta$)

$$\begin{aligned} u(t,x) &= \int_{\mathbb{R}}\int_{\mathbb{R}} \widehat{v}(\xi,\tau)e^{it\tau+ix\xi}d\xi d\tau \\ &= \int_{\mathbb{R}}\int_{\mathbb{R}} e^{it\xi^3}\widehat{v}(\tau+\xi^3,\xi)e^{it\tau+ix\xi}d\xi d\tau \\ &= \int_{\mathbb{R}} e^{it\tau}W_t\psi_\tau d\tau, \end{aligned}$$

where $\widehat{\psi_\tau}(\xi) = \widehat{v}(\tau+\xi^3,\xi)$. Therefore, for each $|t| \leq \delta$

$$\begin{aligned} \|u\|_{H^s_x} &\leq \int_{\mathbb{R}} \|W_t\psi_\tau\|_{H^s}d\tau = \int_{\mathbb{R}} \|\psi_\tau\|_{H^s}d\tau \\ &= \int_{\mathbb{R}} \left\|\widehat{v}(\tau+\xi^3,\xi)\langle\xi\rangle^s\right\|_{L^2_\xi}d\tau \\ &\leq \|\langle\tau\rangle^{-b}\|_{L^2_\tau}\left\|\widehat{v}(\tau+\xi^3,\xi)\langle\xi\rangle^s\langle\tau\rangle^b\right\|_{L^2_\xi L^2_\tau} \\ &\lesssim \|v\|_{X^{s,b}}. \end{aligned}$$

Taking infimum over all such v, we have

$$\|u\|_{L^\infty_{|t|\le\delta} H^s_x} \lesssim \|u\|_{X^{s,b}_\delta}.$$

Continuity in t follows from this, the continuity of the linear group, and the dominated convergence theorem. □

The following lemma describes the behavior of the linear group with respect to $X^{s,b}_\delta$ norm:

Lemma 3.10 *For $0 < \delta \le 1$, $s, b \in \mathbb{R}$, we have*

$$\|\eta(t) W_t g\|_{X^{s,b}_\delta} \lesssim \|g\|_{H^s}.$$

Proof Using the definition of the $X^{s,b}$ norm, we have

$$\|\eta(t) W_t g\|_{X^{s,b}_\delta} \le \|W_t \eta(t) g\|_{X^{s,b}}$$
$$= \|W_{-t} W_t \eta(t) g\|_{H^s_x H^b_t} = \|\eta\|_{H^b}\|g\|_{H^s} \lesssim \|g\|_{H^s}.$$

The first inequality follows from the definition of the restricted norm and the fact that W_t and $\eta(t)$ commute. □

To describe the behavior of the nonlinear part of Γ under the $X^{s,b}_\delta$ norm, we need several lemmas.

Lemma 3.11 *For any $-\frac{1}{2} < b' < b < \frac{1}{2}$ and $s \in \mathbb{R}$, we have*

$$\|u\|_{X^{s,b'}_\delta} \lesssim \delta^{b-b'}\|u\|_{X^{s,b}_\delta}.$$

Proof We will give the proof for $0 \le b' < b < 1/2$. By duality, this implies the inequality for $-1/2 < b' < b \le 0$ as follows

$$\|u\|_{X^{s,b'}_\delta} = \sup_{\|g\|_{X^{-s,-b'}_\delta}=1} \left|\int_{\mathbb{R}^2} ug\, dxdt\right|$$
$$\le \sup_{\|g\|_{X^{-s,-b'}_\delta}=1} \|u\|_{X^{s,b}_\delta}\|g\|_{X^{-s,-b}_\delta} \lesssim \delta^{b-b'}\|u\|_{X^{s,b}_\delta}.$$

By combining these two inequalities, we get the full range.

To obtain the inequality for $0 \le b' < b < 1/2$, first note that by replacing u with $J^s u$ we can assume that $s = 0$. Second, by definition of the restricted norm it suffices to prove that (by taking infimum over $\tilde{u}$)

$$\|\eta(t/\delta)\tilde{u}\|_{X^{0,b'}} \lesssim \delta^{b-b'}\|\tilde{u}\|_{X^{0,b}}.$$

Suppresing the $\tilde{u}$ notation, and using (3.17), we have

$$\|\eta(t/\delta)u\|_{X^{0,b'}} = \|\eta(t/\delta)W_{-t}u\|_{L^2_x H^{b'}_t} = \left\|\left\|\eta(t/\delta)W_{-t}u(t,\widehat{\xi})\right\|_{H^{b'}_t}\right\|_{L^2_\xi}.$$

Therefore, it suffices to prove that

$$\|\eta(t/\delta)f(t)\|_{H^{b'}} \lesssim \delta^{b-b'}\|f\|_{H^b}.$$

Using fractional Leibniz rule (Lemma 1.10) with

$$\frac{1}{p_1} = b - b', \quad \frac{1}{p_2} = b,$$

$$\frac{1}{q_1} = \frac{1}{2} + b' - b, \quad \frac{1}{q_2} = \frac{1}{2} - b,$$

we obtain

$$\begin{aligned}\|\eta(t/\delta)f(t)\|_{H^{b'}} &\lesssim \|\eta(t/\delta)\|_{L^{p_1}}\|J^{b'}f\|_{L^{q_1}} + \|f\|_{L^{q_2}}\|J^{b'}\eta(t/\delta)\|_{L^{p_2}}\\ &\lesssim \|f\|_{H^b}\left(\|\eta(t/\delta)\|_{L^{p_1}} + \|J^{b'}\eta(t/\delta)\|_{L^{p_2}}\right)\\ &\lesssim \|f\|_{H^b}\left(\delta^{b-b'} + \|\eta(t/\delta)\|_{H^{\frac{1}{2}-b+b'}}\right)\\ &\lesssim \delta^{b-b'}\|f\|_{H^b}.\end{aligned}$$

In the second and third inequalities, we used the Sobolev embedding theorem, and scaling in the last two inequalities. □

Lemma 3.12 *Let $-\frac{1}{2} < b' \leq 0$ and $b = b' + 1$. Then*

$$\left\|\eta(t)\int_0^t W_{t-s}F(s)ds\right\|_{X^{s,b}_\delta} \lesssim \|F\|_{X^{s,b'}_\delta}.$$

Proof As before, it suffices to prove the statement with $X^{s,b}$ norms. Note that

$$\left\|\eta(t)\int_0^t W_{t-s}F(s)ds\right\|_{X^{s,b}} = \left\|\eta(t)\int_0^t W_{-s}F(s)ds\right\|_{H^s_xH^b_t}.$$

Therefore, using (3.17) as in the proof of Lemma 3.11, it suffices to prove that

$$\left\|\eta(t)\int_0^t f(s)ds\right\|_{H^b_t} \lesssim \|f\|_{H^{b'}_t}. \tag{3.18}$$

Writing

$$\int_0^t f(s)ds = \int_{\mathbb{R}} \chi_{[0,t]}(s)f(s)ds =$$

$$= \int_{\mathbb{R}} \mathcal{F}^{-1}(\chi_{[0,t]})(z)\widehat{f}(z)dz = \int_{\mathbb{R}} \frac{e^{izt}-1}{iz}\widehat{f}(z)dz,$$

we see that the Fourier transform of the function inside the norm of the left-hand side of (3.18) is

$$\int_{\mathbb{R}} \frac{\widehat{\eta}(\tau - z) - \widehat{\eta}(\tau)}{iz} \widehat{f}(z)dz$$
$$= \int_{|z|<1} \frac{\widehat{\eta}(\tau - z) - \widehat{\eta}(\tau)}{iz} \widehat{f}(z)dz + \int_{|z|>1} \frac{\widehat{\eta}(\tau - z) - \widehat{\eta}(\tau)}{iz} \widehat{f}(z)dz$$
$$=: I(\tau) + II(\tau).$$

For the contribution of the first integral, we use the mean value theorem to get

$$\left\|\langle\tau\rangle^b I(\tau)\right\|_{L^2} \lesssim \left\| \int_{|z|<1} \langle\tau\rangle^b \sup_{|\tau'-\tau|<1} |\widehat{\eta}'(\tau')|\,|\widehat{f}(z)|dz \right\|_{L^2}$$
$$= \left\|\langle\tau\rangle^b \sup_{|\tau'-\tau|<1} |\widehat{\eta}'(\tau')|\right\|_{L^2} \int_{|z|<1} |\widehat{f}(z)|dz$$
$$\lesssim \sqrt{\int_{|z|<1} |\widehat{f}(z)|^2 dz} \lesssim \|f\|_{H^{b'}}.$$

For the contribution of the second integral, we use the inequality

$$\langle\tau\rangle^b \lesssim \langle\tau - z\rangle^b \langle z\rangle^b,$$

and Minkowski's inequality to get

$$\left\|\langle\tau\rangle^b II(\tau)\right\|_{L^2} \le \left\| \langle\tau\rangle^b \int_{\mathbb{R}} \frac{|\widehat{\eta}(\tau - z)| + |\widehat{\eta}(\tau)|}{\langle z\rangle} \left|\widehat{f}(z)\right| dz \right\|_{L^2}$$
$$\lesssim \left\| \int_{\mathbb{R}} \left(\frac{\langle\tau - z\rangle^b |\widehat{\eta}(\tau - z)|}{\langle z\rangle^{1-b}} + \frac{\langle\tau\rangle^b |\widehat{\eta}(\tau)|}{\langle z\rangle} \right) \left|\widehat{f}(z)\right| dz \right\|_{L^2}.$$

By Young's and Cauchy–Schwarz inequalities, we bound this by

$$\left\|\langle\tau\rangle^b \widehat{\eta}\right\|_{L^1} \left\|\langle z\rangle^{b-1} \widehat{f}\right\|_{L^2} + \left\|\langle\tau\rangle^b \widehat{\eta}\right\|_{L^2} \left\|\langle z\rangle^{-1} \widehat{f}\right\|_{L^1}$$
$$\lesssim \left\|\langle z\rangle^{b'} \widehat{f}\right\|_{L^2} + \left\|\langle z\rangle^{b'} \widehat{f}\right\|_{L^2} \left\|\langle z\rangle^{-1-b'}\right\|_{L^2} \lesssim \|f\|_{H^{b'}}.$$

The last inequality follows from the fact that $-1 - b' < -1/2$. □

Remark 3.13 Note that for $b = 1/2$, the proof above implies that

$$\left\|\eta(t) \int_0^t f(s)ds\right\|_{H^{1/2}} \lesssim \|f\|_{H^{-1/2}} + \left\|\langle z\rangle^{-1} \widehat{f}\right\|_{L^1}. \tag{3.19}$$

Lemma 3.14 *For any $b > \frac{1}{2}$ and $b_1 \ge 1/4$ we have*

$$\|\partial_x(uv)\|_{X_\delta^{0,-b_1}} \lesssim \|u\|_{X_\delta^{0,b}} \|v\|_{X_\delta^{0,b}}.$$

Proof Once again, we can ignore δ dependence and work with $X^{s,b}$ norms.
By duality, it suffices to prove that

$$\left|\int_{\mathbb{R}^2} g\partial_x(uv)dxdt\right| \lesssim \|u\|_{X^{0,b}}\|v\|_{X^{0,b}}\|g\|_{X^{0,b_1}}.$$

Using the Fourier multiplication formula (1.3), and renaming the variables, we write the left-hand side as

$$\left|\int_{\mathbb{R}^2} \xi\widehat{uv}(\tau,\xi)\widehat{g}(-\tau,-\xi)d\tau d\xi\right|$$
$$= \left|\int_{\mathbb{R}^4} \xi\widehat{u}(\tau_1,\xi_1)\widehat{v}(\tau-\tau_1,\xi-\xi_1)\widehat{g}(-\tau,-\xi)d\xi d\xi_1 d\tau d\tau_1\right|.$$

Using the notation

$$f_1(\tau,\xi) = |\widehat{u}(\tau,\xi)|\langle\tau-\xi^3\rangle^b,$$
$$f_2(\tau,\xi) = |\widehat{v}(\tau,\xi)|\langle\tau-\xi^3\rangle^b,$$
$$f_3(\tau,\xi) = |\widehat{g}(-\tau,-\xi)|\langle\tau-\xi^3\rangle^{b_1},$$

it suffices to prove that

$$\int_{\mathbb{R}^4} \frac{|\xi|f_1(\tau_1,\xi_1)f_2(\tau-\tau_1,\xi-\xi_1)f_3(\tau,\xi)}{\langle\tau_1-\xi_1^3\rangle^b\langle\tau-\tau_1-(\xi-\xi_1)^3\rangle^b\langle\tau-\xi^3\rangle^{b_1}}d\xi d\xi_1 d\tau d\tau_1 \lesssim \prod_{i=1}^{3}\|f_i\|_2. \quad (3.20)$$

We claim that

$$\sup_{\xi,\tau}\frac{|\xi|^2}{\langle\tau-\xi^3\rangle^{2b_1}}\int_{\mathbb{R}^2}\frac{1}{\langle\tau_1-\xi_1^3\rangle^{2b}\langle\tau-\tau_1-(\xi-\xi_1)^3\rangle^{2b}}d\xi_1 d\tau_1 \lesssim 1. \quad (3.21)$$

By using Cauchy–Schwarz inequality and the claim, we estimate the left-hand side of (3.20) by

$$\left(\int_{\mathbb{R}^4} f_1^2(\tau_1,\xi_1)f_2^2(\tau-\tau_1,\xi-\xi_1)d\xi d\xi_1 d\tau d\tau_1\right)^{1/2}$$
$$\times\left(\int_{\mathbb{R}^4}\frac{|\xi|^2 f_3^2(\tau,\xi)}{\langle\tau_1-\xi_1^3\rangle^{2b}\langle\tau-\tau_1-(\xi-\xi_1)^3\rangle^{2b}\langle\tau-\xi^3\rangle^{2b_1}}d\xi d\xi_1 d\tau d\tau_1\right)^{1/2}$$
$$\lesssim \|f_1\|_2\|f_2\|_2\left(\int_{\mathbb{R}^2} f_3^2(\tau,\xi)d\xi d\tau\right)^{1/2} = \prod_{i=1}^{3}\|f_i\|_2.$$

It remains to prove (3.21). Using Exercise 3.12 in the τ_1 integral (for $b > \frac{1}{2}$), we bound the left-hand side of (3.21) by

$$\sup_{\xi,\tau}\frac{|\xi|^2}{\langle\tau-\xi^3\rangle^{2b_1}}\int_{\mathbb{R}}\frac{1}{\langle\tau-\xi_1^3-(\xi-\xi_1)^3\rangle^{2b}}d\xi_1. \quad (3.22)$$

Let $x = \tau - \xi_1^3 - (\xi - \xi_1)^3$. Using

$$\xi_1 = \frac{3\xi^2 \pm \sqrt{3\xi(4\tau - \xi^3 - 4x)}}{6\xi},$$

we obtain

$$dx = (3\xi^2 - 6\xi\xi_1)d\xi = \pm\sqrt{3\xi(4\tau - \xi^3 - 4x)}\, d\xi_1.$$

Therefore, we can estimate (3.22) by

$$\sup_{\xi,\tau} \frac{|\xi|^2}{\langle \tau - \xi^3 \rangle^{2b_1}} \int_{\mathbb{R}} \frac{1}{\langle x \rangle^{2b} \sqrt{|\xi|}\sqrt{|4\tau - \xi^3 - 4x|}} dx.$$

Using Exercise 3.13 (for $b > \frac{1}{2}$), we obtain

$$\sup_{\xi,\tau} \frac{|\xi|^{3/2}}{\langle \tau - \xi^3 \rangle^{2b_1} \langle 4\tau - \xi^3 \rangle^{1/2}} \lesssim 1.$$

The last inequality holds since $b_1 \geq \frac{1}{4}$. □

We now run the contraction argument in $B_\delta \subset X_\delta^{0,b}$ (with $b > \frac{1}{2}$ and δ sufficiently small) for the operator

$$\Gamma u = \eta(t) W_t g - \eta(t) \int_0^t W_{t-s}(uu_x) ds.$$

Using the bounds in Lemma 3.10 and in Lemma 3.12, we have

$$\|\Gamma u\|_{X_\delta^{0,b}} \lesssim \|g\|_{L^2} + \|uu_x\|_{X_\delta^{0,b-1}}.$$

Now, using Lemma 3.11 (with $b_1 \geq \frac{1}{4}$) and then Lemma 3.14, we obtain

$$\begin{aligned} \|\Gamma u\|_{X_\delta^{0,b}} &\lesssim \|g\|_{L^2} + \delta^{1-b-b_1} \|uu_x\|_{X_\delta^{0,-b_1}} \\ &\lesssim \|g\|_{L^2} + \delta^{1-b-b_1} \|u\|_{X_\delta^{0,b}}^2. \end{aligned}$$

Similar estimates hold for the differences. Therefore, Γ is a contraction on B_δ for any $b > \frac{1}{2}$, $b_1 \geq \frac{1}{4}$, $1 - b - b_1 > 0$ and for $\delta = \delta(\|g\|_{L^2}, b, b_1)$ sufficiently small.

3.3.2 Low regularity solutions of KdV on the torus

In this section, we obtain the local wellposedness of the KdV equation on the torus

$$\begin{cases} u_t + u_{xxx} + uu_x = 0, & t \in \mathbb{R},\ x \in \mathbb{T} \\ u(0, x) = g(x) \in H^s(\mathbb{T}), \end{cases} \tag{3.23}$$

for $s > -\frac{1}{2}$. The natural candidate for the local wellposedness theory is the $X^{s,b}$ space on the torus defined via the norm

$$\|u\|_{X^{s,b}} = \left\|W_{-t}u\right\|_{H^s_x H^b_t(\mathbb{T}\times\mathbb{R})} = \left\|\langle k\rangle^s \langle \tau - k^3\rangle^b \widehat{u}(\tau, k)\right\|_{\ell^2_k L^2_\tau}.$$

Unlike the case of the real line, one cannot work with the $X^{s,b}$ spaces with $b > \frac{1}{2}$. Indeed, it was proved in [91] that the claim of Lemma 3.14 does not hold in the periodic setting for any s and for any $b \neq \frac{1}{2}$, even when $b_1 = 1 - b$. However, the space $X^{s,\frac{1}{2}}$ does not embed into $C^0_t H^s_x(\mathbb{R} \times \mathbb{T})$. Bourgain [16] overcame this difficulty by introducing a modified space, Y^s, based on the idea that the Fourier transform of an L^1 function is continuous. We define the space Y^s via the norm

$$\|u\|_{Y^s} = \|u\|_{X^{s,1/2}} + \|\langle k\rangle^s \widehat{u}(\tau, k)\|_{\ell^2_k L^1_\tau}.$$

The restricted space Y^s_δ is defined accordingly.

Note that for each t

$$\begin{aligned}\|u(t,\cdot)\|^2_{H^s} &= \sum_k |\widehat{u(t,k)}|^2 \langle k\rangle^{2s} \\ &= \sum_k \left|\int_{\mathbb{R}} \widehat{u}(\tau, k) e^{it\tau} d\tau\right|^2 \langle k\rangle^{2s} \\ &\leq \sum_k \left(\int_{\mathbb{R}} |\widehat{u}(\tau, k)| d\tau\right)^2 \langle k\rangle^{2s} \leq \|u\|^2_{Y^s}.\end{aligned}$$

Continuity in t follows by the dominated convergence theorem. Thus, Y^s embeds into $C^0_t H^s_x(\mathbb{R} \times \mathbb{T})$.

Recall from Section 3.1 that the mean $I_1(t) = \int_{\mathbb{T}} u(t,x)dx$ is conserved. In this section, we consider mean-zero solutions of the KdV equation. This assumption can be justified as follows (we note that on $\mathbb{R}$ this idea fails): Let u solve (3.23) with data g. Now set $v(t,x) = u(t,x) + c$ and observe that v solves

$$v_t + v_{xxx} - cv_x + vv_x = 0$$

with $v(0,x) = g(x) + c$. We choose

$$c = -\frac{1}{2\pi}\int_0^{2\pi} g(x)dx$$

so that v is mean-zero. Although v does not solve the original KdV anymore, the methods we outline apply to the new equation step by step. The only difference is that now the multiplier of the linear group is $k^3 + ck$ instead k^3. Notice that in all calculations that follow this replacement changes nothing.

We also note that the wellposedness result we establish implies Lipcshitz dependence on initial data on sets of the form

$$\left\{g \in H^s : \|g\|_{H^s} \le M, \int_{\mathbb{T}} g(x)dx = c\right\}$$

for any fixed M and c. It is known that without the fixed mean condition Lipcshitz dependence fails for any s; see Molinet [106]. We demonstrate this in Section 3.6 below.

We now present suitable versions of the lemmas in the previous section for the Y^s norms. As before, we ignore the δ dependence in the proofs.

Lemma 3.15 *For any s real*

$$\|\eta(t)W_t g\|_{Y^s_\delta} \lesssim \|g\|_{H^s}.$$

Proof The proof for the $X^{s,\frac{1}{2}}$ part is identical to the one given in Lemma 3.10. Therefore, the following calculation yields the claim

$$\|\mathcal{F}(\eta W_t g)(\tau,k)\|_{\ell^2_k L^1_\tau} = \left\|\widehat{\eta}(\tau + k^3)\widehat{g}(k)\langle k\rangle^s\right\|_{\ell^2_k L^1_\tau} \lesssim \|g\|_{H^s}.$$

□

For the one derivative gain in Lemma 3.12, we need to define the space Z^s via the norm

$$\|u\|_{Z^s} = \|u\|_{X^{s,-1/2}} + \left\|\langle k\rangle^s \left\langle\tau - k^3\right\rangle^{-1} \widehat{u}(\tau,k)\right\|_{\ell^2_k L^1_\tau}.$$

Again, Z^s_δ is defined accordingly.

Lemma 3.16 *We have*

$$\left\|\eta(t)\int_0^t W_{t-s}F(s)ds\right\|_{Y^s_\delta} \lesssim \|F\|_{Z^s_\delta}.$$

Proof We first estimate the $X^{s,1/2}$ part of the norm. Using (3.17), we have

$$\begin{aligned}\left\|\eta(t)\int_0^t W_{t-s}F(s)ds\right\|_{X^{s,1/2}} &= \left\|\eta(t)\int_0^t W_{-s}F(s)ds\right\|_{H^s_x H^{\frac{1}{2}}_t} \\ &= \left\|\left\|\eta(t)\int_0^t [W_{-s}F(s)]\,\widehat{(k)}ds\right\|_{H^{\frac{1}{2}}_t}\langle k\rangle^s\right\|_{\ell^2_k}.\end{aligned}$$

Recalling (3.19)

$$\left\|\eta(t)\int_0^t f(s)ds\right\|_{H^{1/2}} \lesssim \|f\|_{H^{-1/2}} + \left\|\langle z\rangle^{-1}\widehat{f}\right\|_{L^1},$$

we estimate this by

$$\left\| \left\| [W_{-t}F(t)]\widehat{(k)} \right\|_{H_t^{-1/2}} \langle k \rangle^s \right\|_{\ell_k^2} + \left\| \widehat{W_{-t}F}(t)(k,\tau)\langle k \rangle^s \langle \tau \rangle^{-1} \right\|_{\ell_k^2 L_\tau^1} = \|F\|_{Z^s}.$$

To estimate the other part of the Y^s norm, define $\mathcal{D}(x,t) = \eta(t) \int_0^t W_{t-s}F(s)ds$. Recall from the proof of Lemma 3.12 that

$$\widehat{\mathcal{D}}(k,\tau) = \int_{\mathbb{R}} \frac{\widehat{\eta}(\tau - z - k^3) - \widehat{\eta}(\tau - k^3)}{iz} \widehat{F}(z + k^3, k) dz.$$

Using this, we estimate

$$\begin{aligned}
&\left\| \langle k \rangle^s \widehat{\mathcal{D}}(k,\tau) \right\|_{\ell_k^2 L_\tau^1} \\
&\quad \le \left\| \int_{\mathbb{R}} \left\| \frac{\widehat{\eta}(\tau - z - k^3) - \widehat{\eta}(\tau - k^3)}{iz} \right\|_{L_\tau^1} \left| \widehat{F}(z + k^3, k) \right| \langle k \rangle^s dz \right\|_{\ell_k^2} \\
&\qquad\qquad \lesssim \left\| \int_{\mathbb{R}} \langle z \rangle^{-1} |\widehat{F}(z + k^3, k)| \langle k \rangle^s dz \right\|_{\ell_k^2} \le \|F\|_{Z^s}.
\end{aligned}$$

We obtained the second line by considering the cases $|z| < 1$ and $|z| > 1$ separately. In the former case, we used the mean value theorem; see the proof of Lemma 3.12. □

The following is a simple corallary of the Strichartz estimates we have in Theorem 2.9:

Corollary 3.17 *Fix $\delta \lesssim 1$ and $b > \frac{1}{2}$. For any space-time function u, we have*

(i) $\|u\|_{L^4_{x\in\mathbb{T},\, t\in[-\delta,\delta]}} \lesssim \|u\|_{X_\delta^{0,b}}$,

(ii) $\|u\|_{L^6_{x\in\mathbb{T},\, t\in[-\delta,\delta]}} \lesssim \|u\|_{X_\delta^{\epsilon,b}}$, *for any $\epsilon > 0$.*

Proof By Fourier inversion and then a change of variable, we have

$$\begin{aligned}
u(t,x) &= \sum_{k\in\mathbb{Z}} \int_{\mathbb{R}} \widehat{u}(\tau,k) e^{it\tau + ixk} d\tau \\
&= \sum_{k\in\mathbb{Z}} \int_{\mathbb{R}} e^{itk^3} \widehat{u}(\tau + k^3, k) e^{it\tau + ixk} d\tau \\
&= \int_{\mathbb{R}} e^{it\tau} W_t \psi_\tau d\tau,
\end{aligned}$$

where $\widehat{\psi_\tau}(k) = \widehat{u}(\tau + k^3, k)$. Therefore

$$\begin{aligned}\|u\|_{L^4_{x\in\mathbb{T},\,t\in[-\delta,\delta]}} &\leq \int_{\mathbb{R}} \|W_t\psi_\tau\|_{L^4_{x,t\in\mathbb{T}}} d\tau \lesssim \int_{\mathbb{R}} \|\psi_\tau\|_{L^2} d\tau \\ &= \int_{\mathbb{R}} \big\|\widehat{u}(\tau + k^3, k)\big\|_{\ell^2_k} d\tau \\ &\leq \|\langle\tau\rangle^{-b}\|_{L^2_\tau} \big\|\widehat{u}(\tau + k^3, k)\langle\tau\rangle^b\big\|_{\ell^2_k L^2_\tau} \\ &\lesssim \|u\|_{X^{0,b}}.\end{aligned}$$

The proof of (ii) is similar. □

By Lemma 3.16, to close the fixed point argument, we need to estimate the Z^s norm of the nonlinearity, uu_x, by the Y^s norm of the solution. For this, we need Bourgain's refinement (see [15, 16]) of the L^4 Strichartz estimate, part (i) of Corollary 3.17. This refinement allows us to extract a power of δ in the wellposedness proof; also see the discussion in Tao [143].

Theorem 3.18 *For any space-time function u*

$$\|u\|_{L^4_{x\in\mathbb{T},\,t\in\mathbb{R}}} \lesssim \|u\|_{X^{0,1/3}}.$$

Proof Let $u = \sum_{m=0}^{\infty} u_{2^m}$, where

$$\widehat{u_{2^m}}(\tau, k) = \widehat{u}(\tau, k)\chi_{2^m \leq \langle\tau - k^3\rangle < 2^{m+1}}.$$

Note that by Plancherel

$$\|u\|^2_{X^{0,1/3}} \approx \sum_{m=0}^{\infty} 2^{2m/3} \|u_{2^m}\|^2_{L^2_{x,t}}. \tag{3.24}$$

We write

$$\|u\|^2_{L^4_{x,t}} = \big\|u^2\big\|_{L^2_{x,t}} \leq 2 \sum_{m\leq m'} \big\|u_{2^m}u_{2^{m'}}\big\|_{L^2_{x,t}} = 2 \sum_{m,n\geq 0} \|u_{2^m}u_{2^{m+n}}\|_{L^2_{x,t}}.$$

Note that

$$\begin{aligned}\|u_{2^m}u_{2^{m+n}}\|_{L^2_{x,t}} &= \big\|\widehat{u_{2^m}} * \widehat{u_{2^{m+n}}}\big\|_{\ell^2_k L^2_\tau} \\ &= \Bigg\|\sum_{k_1\in\mathbb{Z}} \int_{\mathbb{R}} \widehat{u_{2^m}}(\tau_1, k_1)\widehat{u_{2^{m+n}}}(\tau - \tau_1, k - k_1)d\tau_1\Bigg\|_{\ell^2_k L^2_\tau}.\end{aligned} \tag{3.25}$$

We estimate this separately in the range $|k| \leq 2^a$ and $|k| > 2^a$, where a will be determined later. In the former case, for each $|k| \leq 2^a$, we put the L^2_τ norm inside the sum and apply Young's inequality to obtain

$$\left\|\widehat{u_{2^m}} * \widehat{u_{2^{m+n}}}\right\|_{L^2_\tau} \leq \sum_{k_1\in\mathbb{Z}} \left\| \int_{\mathbb{R}} \widehat{u_{2^m}}(\tau_1, k_1)\widehat{u_{2^{m+n}}}(\tau - \tau_1, k - k_1)d\tau_1 \right\|_{L^2_\tau}$$

$$\leq \sum_{k_1\in\mathbb{Z}} \left\|\widehat{u_{2^m}}(\cdot, k_1)\right\|_{L^1} \left\|\widehat{u_{2^{m+n}}}(\cdot, k - k_1)\right\|_{L^2}. \quad (3.26)$$

Using the Cauchy–Schwarz inequality in the L^1 norm by taking into account the support condition on τ, we estimate this sum as

$$(3.26) \lesssim 2^{\frac{m}{2}} \sum_{k_1\in\mathbb{Z}} \left\|\widehat{u_{2^m}}(\cdot, k_1)\right\|_{L^2} \left\|\widehat{u_{2^{m+n}}}(\cdot, k - k_1)\right\|_{L^2}$$

$$\leq 2^{\frac{m}{2}} \left(\sum_{k_1\in\mathbb{Z}} \left\|\widehat{u_{2^m}}(\cdot, k_1)\right\|^2_{L^2}\right)^{1/2} \left(\sum_{k_1\in\mathbb{Z}} \left\|\widehat{u_{2^{m+n}}}(\cdot, k - k_1)\right\|^2_{L^2}\right)^{1/2}$$

$$= 2^{\frac{m}{2}} \|u_{2^m}\|_{L^2_{x,t}} \|u_{2^{m+n}}\|_{L^2_{x,t}}.$$

Therefore, taking the $\ell^2_{|k|\leq 2^a}$ norm, we obtain

$$\left\|\widehat{u_{2^m}} * \widehat{u_{2^{m+n}}}\right\|_{\ell^2_{|k|\leq 2^a} L^2_\tau} \lesssim 2^{\frac{a+m}{2}} \|u_{2^m}\|_{L^2_{x,t}} \|u_{2^{m+n}}\|_{L^2_{x,t}}. \quad (3.27)$$

In the latter case, by using the Cauchy–Schwarz inequality in the k_1, τ_1 variables, we have

$$\left\|\widehat{u_{2^m}} * \widehat{u_{2^{m+n}}}\right\|_{\ell^2_{|k|>2^a} L^2_\tau} \leq$$

$$\left\| \left(\sum_{k_1\in\mathbb{Z}} \int_{\mathbb{R}} \left|\widehat{u_{2^m}}(\tau_1, k_1)\right|^2 \left|\widehat{u_{2^{m+n}}}(\tau - \tau_1, k - k_1)\right|^2 d\tau_1 \right)^{1/2} (\chi_m * \chi_{m+n}(\tau, k))^{1/2} \right\|_{\ell^2_{|k|>2^a} L^2_\tau},$$

where

$$\chi_m(\tau, k) := \chi_{2^m \leq \langle \tau - k^3\rangle < 2^{m+1}}.$$

Taking the supremum of the convolution outside the norm, we obtain

$$\left\|\widehat{u_{2^m}} * \widehat{u_{2^{m+n}}}\right\|_{\ell^2_{|k|>2^a} L^2_\tau}$$

$$\leq \|\chi_m * \chi_{m+n}\|^{1/2}_{\ell^\infty_{|k|>2^a} L^\infty_\tau} \left\| \left(\sum_{k_1\in\mathbb{Z}} \int_{\mathbb{R}} \left|\widehat{u_{2^m}}(\tau_1, k_1)\right|^2 \left|\widehat{u_{2^{m+n}}}(\tau - \tau_1, k - k_1)\right|^2 d\tau_1 \right)^{1/2} \right\|_{\ell^2_k L^2_\tau}$$

$$= \|\chi_m * \chi_{m+n}\|^{1/2}_{\ell^\infty_{|k|>2^a} L^\infty_\tau} \|u_{2^m}\|_{L^2_{x,t}} \|u_{2^{m+n}}\|_{L^2_{x,t}}.$$

To estimate the convolution, we write for fixed $|k| > 2^a$ and τ

$$\chi_m * \chi_{m+n}(\tau, k) = \sum_{k_1\in\mathbb{Z}} \int_{\mathbb{R}} \chi_m(\tau_1, k_1)\chi_{m+n}(\tau - \tau_1, k - k_1)d\tau_1.$$

By the support condition on χ_m and χ_{m+n}, we have

$$\tau_1 = k_1^3 + O(2^m),$$
$$\tau - \tau_1 = (k-k_1)^3 + O(2^{m+n}).$$

Therefore, for each fixed k_1, the τ_1 integral is $O(2^m)$. To calculate the number of k_1s for which the integral is nonzero, note that

$$\tau = k_1^3 + (k-k_1)^3 + O(2^{m+n}),$$

and hence

$$k^2 - 3k_1k + 3k_1^2 = \frac{\tau}{k} + O(2^{m+n-a}).$$

This implies that

$$3(k_1 - k/2)^2 = \frac{\tau}{k} - \frac{k^2}{4} + O(2^{m+n-a}).$$

Therefore, k_1 takes $O(2^{\frac{m+n-a}{2}})$ values, which implies

$$\|\chi_m * \chi_{m+n}\|_{\ell^\infty_{|k|>2^a}L^\infty_\tau} \lesssim 2^{\frac{3m+n-a}{2}}.$$

Combining these bounds, we obtain

$$\left\|\widehat{u_{2^m}} * \widehat{u_{2^{m+n}}}\right\|_{\ell^2_{|k|>2^a}L^2_\tau} \lesssim 2^{\frac{3m+n-a}{4}} \|u_{2^m}\|_{L^2_{x,t}} \|u_{2^{m+n}}\|_{L^2_{x,t}}. \tag{3.28}$$

Using (3.27) and (3.28) with $a = \frac{m+n}{3}$, we obtain

$$\left\|\widehat{u_{2^m}} * \widehat{u_{2^{m+n}}}\right\|_{\ell^2_kL^2_\tau} \lesssim 2^{\frac{4m+n}{6}} \|u_{2^m}\|_{L^2_{x,t}} \|u_{2^{m+n}}\|_{L^2_{x,t}}.$$

Finally, we estimate

$$\begin{aligned}\|u\|^2_{L^4_{x,t}} &\le 2\sum_{m,n\ge 0} \|u_{2^m}u_{2^{m+n}}\|_{L^2} = 2\sum_{m,n\ge 0}\left\|\widehat{u_{2^m}} * \widehat{u_{2^{m+n}}}\right\|_{\ell^2_kL^2_\tau}\\ &\lesssim \sum_{n\ge 0} 2^{-\frac{n}{6}} \sum_{m\ge 0} 2^{\frac{m}{3}} \|u_{2^m}\|_{L^2_{x,t}} 2^{\frac{m+n}{3}} \|u_{2^{m+n}}\|_{L^2_{x,t}}.\end{aligned}$$

By the Cauchy–Schwarz inequality in the m sum and using (3.24), we conclude that

$$\|u\|^2_{L^4_{x,t}} \lesssim \|u\|^2_{X^{0,\frac{1}{3}}} \sum_{n\ge 0} 2^{-\frac{n}{6}} \lesssim \|u\|^2_{X^{0,\frac{1}{3}}}.$$

□

Theorem 3.19 *Assume that u and v are space-time functions of mean-zero for each t, then for $s > -\frac{1}{2}$ we have*

$$\|\partial_x(uv)\|_{Z^s_\delta} \lesssim \|u\|_{X^{s,\frac{1}{2}}_\delta} \|v\|_{X^{s,\frac{1}{3}}_\delta} + \|u\|_{X^{s,\frac{1}{3}}_\delta} \|v\|_{X^{s,\frac{1}{2}}_\delta}.$$

Proof We will give the proof only for the range $s \in (-\frac{1}{2}, 0]$. The proof is easier for $s > 0$. We start with the first part of the Z^s norm

$$\begin{aligned}\|\partial_x(uv)\|_{X^{s,-\frac{1}{2}}} &= \sup_{\|w\|_{X^{-s,\frac{1}{2}}}=1} \left|\int_{\mathbb{R}^2} w\partial_x(uv)dtdx\right| \\ &= \sup_{\|w\|_{X^{-s,\frac{1}{2}}}=1} \left|\sum_{k_1+k_2+k_3=0} \int_{\tau_1+\tau_2+\tau_3=0} k_3\widehat{u}(\tau_1,k_1)\widehat{v}(\tau_2,k_2)\widehat{w}(\tau_3,k_3)\right|.\end{aligned}$$

Using the notation

$$\begin{aligned}f_1(\tau,k) &= |\widehat{u}(\tau,k)|\langle k\rangle^s\langle\tau-k^3\rangle^{1/2},\\ f_2(\tau,k) &= |\widehat{v}(\tau,k)|\langle k\rangle^s\langle\tau-k^3\rangle^{1/2},\\ f_3(\tau,k) &= |\widehat{w}(\tau,k)|\langle k\rangle^{-s}\langle\tau-k^3\rangle^{1/2},\end{aligned}$$

we estimate the right-hand side by

$$\sum_{k_1+k_2+k_3=0}\int_{\tau_1+\tau_2+\tau_3=0} \frac{\langle k_3\rangle^{1+s} f_1(\tau_1,k_1)f_2(\tau_2,k_2)f_3(\tau_3,k_3)}{\langle k_1\rangle^s\langle k_2\rangle^s\langle\tau_1-k_1^3\rangle^{1/2}\langle\tau_2-k_2^3\rangle^{1/2}\langle\tau_3-k_3^3\rangle^{1/2}}. \tag{3.29}$$

Note that because of the mean-zero assumption, $k_j \neq 0$ in the sum above. We continue by rewriting the multiplier by setting $s = -\rho \in [0, 1/2)$

$$\frac{\langle k_1\rangle^\rho\langle k_2\rangle^\rho\langle k_3\rangle^{1-\rho}}{\langle\tau_1-k_1^3\rangle^{1/2}\langle\tau_2-k_2^3\rangle^{1/2}\langle\tau_3-k_3^3\rangle^{1/2}}.$$

Notice that

$$\begin{aligned}\tau_1 - k_1^3 + \tau_2 - k_2^3 + \tau_3 - k_3^3 &= (k_1+k_2)^3 - k_1^3 - k_2^3 \\ &= 3k_1k_2(k_1+k_2) = -3k_1k_2k_3.\end{aligned}$$

Therefore (using $k_j \neq 0$)

$$\max\Big(\langle\tau_1-k_1^3\rangle, \langle\tau_2-k_2^3\rangle, \langle\tau_3-k_3^3\rangle\Big) \gtrsim \langle k_1\rangle\langle k_2\rangle\langle k_3\rangle. \tag{3.30}$$

Assume that the largest one is $\langle\tau_1 - k_1^3\rangle$, the other cases are similar. The multiplier is estimated by (using $k_3 = -k_1 - k_2$)

$$\frac{\langle k_3\rangle^{\frac{1}{2}-\rho}}{\langle k_1\rangle^{\frac{1}{2}-\rho}\langle k_2\rangle^{\frac{1}{2}-\rho}\langle\tau_2-k_2^3\rangle^{1/2}\langle\tau_3-k_3^3\rangle^{1/2}} \lesssim \frac{1}{\langle\tau_2-k_2^3\rangle^{1/2}\langle\tau_3-k_3^3\rangle^{1/2}}.$$

Using this in (3.29), and ignoring the contribution of similar terms, we obtain

$$\begin{aligned}(3.29) &\lesssim \sum_{k_1+k_2+k_3=0}\int_{\tau_1+\tau_2+\tau_3=0} \frac{f_1(\tau_1,k_1)f_2(\tau_2,k_2)f_3(\tau_3,k_3)}{\langle\tau_2-k_2^3\rangle^{1/2}\langle\tau_3-k_3^3\rangle^{1/2}} \\ &= \int_{\mathbb{T}\times\mathbb{R}} \mathcal{F}^{-1}(f_1)\mathcal{F}^{-1}\left(\frac{f_2}{\langle\tau-k^3\rangle^{1/2}}\right)\mathcal{F}\left(\frac{f_3}{\langle\tau-k^3\rangle^{1/2}}\right)dxdt.\end{aligned}$$

Here we used (space-time) Fourier multiplication formula and the convolution structure. By using Hölder's inequality and then Theorem 3.18, we estimate the integral by

$$\begin{aligned}
&\|\mathcal{F}^{-1}(f_1)\|_{L^2_{x,t}} \left\|\mathcal{F}^{-1}\left(\frac{f_2}{\langle\tau-k^3\rangle^{1/2}}\right)\right\|_{L^4_{x,t}} \left\|\mathcal{F}\left(\frac{f_3}{\langle\tau-k^3\rangle^{\frac{1}{2}}}\right)\right\|_{L^4_{x,t}} \\
&\qquad \lesssim \|f_1\|_{L^2} \left\|\mathcal{F}^{-1}\left(\frac{f_2}{\langle\tau-k^3\rangle^{\frac{1}{2}}}\right)\right\|_{X^{0,\frac{1}{3}}} \left\|\mathcal{F}\left(\frac{f_3}{\langle\tau-k^3\rangle^{\frac{1}{2}}}\right)\right\|_{X^{0,\frac{1}{3}}} \\
&\qquad = \|u\|_{X^{s,\frac{1}{2}}} \|v\|_{X^{s,\frac{1}{3}}} \|w\|_{X^{-s,\frac{1}{3}}} \leq \|u\|_{X^{s,\frac{1}{2}}} \|v\|_{X^{s,\frac{1}{3}}} \|w\|_{X^{-s,\frac{1}{2}}}.
\end{aligned}$$

We continue with the second part of the Z^s norm. Using duality, we write

$$\begin{aligned}
&\left\|\frac{\langle k\rangle^s \widehat{\partial_x(uv)}(\tau,k)}{\langle\tau-k^3\rangle}\right\|_{\ell^2_k L^1_\tau} \\
&\lesssim \sup_{\|w\|_{\ell^2_k L^\infty_\tau}=1} \sum_{k_1+k_2+k_3=0} \int_{\tau_1+\tau_2+\tau_3=0} \frac{\langle k_3\rangle^{1+s}|\widehat{u}(\tau_1,k_1)||\widehat{v}(\tau_2,k_2)||w(\tau_3,k_3)|}{\langle\tau_3-k_3^3\rangle} \\
&= \sup_{\|w\|_{\ell^2_k L^\infty_\tau}=1} \sum_{k_1+k_2+k_3=0} \int_{\tau_1+\tau_2+\tau_3=0} \frac{\langle k_3\rangle^{1+s} f_1(\tau_1,k_1) f_2(\tau_2,k_2)|w(k_3,\tau_3)|}{\langle k_1\rangle^s\langle k_2\rangle^s\langle\tau_1-k_1^3\rangle^{\frac{1}{2}}\langle\tau_2-k_2^3\rangle^{\frac{1}{2}}\langle\tau_3-k_3^3\rangle},
\end{aligned}$$

with f_1 and f_2 as above. By symmetry, we have two cases to consider:

Case 1: $\max(\langle\tau_1-k_1^3\rangle, \langle\tau_2-k_2^3\rangle, \langle\tau_3-k_3^3\rangle) = \langle\tau_1-k_1^3\rangle$.
Using (3.30), the multiplier is bounded by

$$\frac{\langle k_3\rangle^{\frac{1}{2}+s}}{\langle k_1\rangle^{\frac{1}{2}+s}\langle k_2\rangle^{\frac{1}{2}+s}\langle\tau_2-k_2^3\rangle^{\frac{1}{2}}\langle\tau_3-k_3^3\rangle} \lesssim \frac{1}{\langle\tau_2-k_2^3\rangle^{\frac{1}{2}}\langle\tau_3-k_3^3\rangle}.$$

Using this bound as above, we estimate the norm in this case by

$$\begin{aligned}
&\sup_{\|w\|_{\ell^2_k L^\infty_\tau}=1} \|f_1\|_{L^2} \left\|\mathcal{F}^{-1}\left(\frac{f_2}{\langle\tau-k^3\rangle^{1/2}}\right)\right\|_{X^{0,\frac{1}{3}}} \left\|\mathcal{F}\left(\frac{|w|}{\langle\tau-k^3\rangle}\right)\right\|_{X^{0,\frac{1}{3}}} \\
&\qquad = \|u\|_{X^{s,\frac{1}{2}}} \|v\|_{X^{s,\frac{1}{3}}} \sup_{\|w\|_{\ell^2_k L^\infty_\tau}=1} \left\|\frac{w}{\langle\tau-k^3\rangle^{\frac{2}{3}}}\right\|_{\ell^2_k L^2_\tau} \\
&\qquad \lesssim \|u\|_{X^{s,\frac{1}{2}}} \|v\|_{X^{s,\frac{1}{3}}}.
\end{aligned}$$

In the last line, we used Hölder's inequality in the τ variable.

Case 2: $\max(\langle\tau_1-k_1^3\rangle, \langle\tau_2-k_2^3\rangle, \langle\tau_3-k_3^3\rangle) = \langle\tau_3-k_3^3\rangle$.
Using (3.30), we estimate

$$\langle\tau_3-k_3^3\rangle \gtrsim \langle k_1\rangle\langle k_2\rangle\langle k_3\rangle \gtrsim \langle k_3\rangle^2.$$

Therefore, we have

$$\langle \tau_3 - k_3^3 \rangle = \langle \tau_3 - k_3^3 \rangle^{-s} \langle \tau_3 - k_3^3 \rangle^{1+s}$$
$$\gtrsim \langle k_1 \rangle^{-s} \langle k_2 \rangle^{-s} \langle k_3 \rangle^{-s} (\langle \tau_3 - k_3^3 \rangle + \langle k_3 \rangle^2)^{1+s}.$$

Using this, we estimate the multiplier by

$$\frac{\langle k_3 \rangle^{1+2s}}{\langle \tau_1 - k_1^3 \rangle^{\frac{1}{2}} \langle \tau_2 - k_2^3 \rangle^{\frac{1}{2}} (\langle \tau_3 - k_3^3 \rangle + \langle k_3 \rangle^2)^{1+s}}.$$

Using this as above (switching the roles of f_1 and w), we bound the norm by

$$\|u\|_{X^{s,\frac{1}{3}}} \|v\|_{X^{s,\frac{1}{3}}} \sup_{\|w\|_{\ell_k^2 L_\tau^\infty}=1} \left\| \frac{w(k,\tau)\langle k \rangle^{1+2s}}{(\langle \tau - k^3 \rangle + \langle k \rangle^2)^{1+s}} \right\|_{\ell_k^2 L_\tau^2}$$
$$\lesssim \|u\|_{X^{s,\frac{1}{3}}} \|v\|_{X^{s,\frac{1}{3}}} \left\| \frac{\langle k \rangle^{1+2s}}{(\langle \tau - k^3 \rangle + \langle k \rangle^2)^{1+s}} \right\|_{\ell_k^\infty L_\tau^2}$$
$$\lesssim \|u\|_{X^{s,\frac{1}{3}}} \|v\|_{X^{s,\frac{1}{3}}}.$$

The last line follows from the inequality (using $s > -\frac{1}{2}$)

$$\int_{\mathbb{R}} \frac{1}{(|\tau| + \langle k \rangle^2)^{2+2s}} d\tau \lesssim \frac{1}{\langle k \rangle^{2(1+2s)}}.$$

□

The corollary below follows immediately from Theorem 3.19 and Lemma 3.11 (which holds also in the periodic case).

Corollary 3.20 *Let $\delta \in (0,1)$. Assume that u is a space-time function of mean-zero for each t, then for $s > -\frac{1}{2}$ we have*

$$\left\| \partial_x(u^2) \right\|_{Z_\delta^s} \lesssim \delta^{\frac{1}{6}-} \|u\|^2_{X_\delta^{s,1/2}} \leq \delta^{\frac{1}{6}-} \|u\|^2_{Y_\delta^s}.$$

Using Lemma 3.15, Lemma 3.16, and Corollary 3.20, we have

$$\|\Gamma u\|_{Y_\delta^s} \lesssim \|g\|_{H^s} + \delta^{\frac{1}{6}-} \|u\|^2_{Y_\delta^s},$$

where

$$\Gamma u = \eta(t) W_t g - \eta(t) \int_0^t W_{t-s}(u u_x) ds.$$

Thus, one can close the contraction argument as before.

3.3.3 Forced and damped KdV with a potential

In this section, we extend the local wellposedness theory of the previous section to the forced and damped KdV equation with a potential on the torus

$$\begin{cases} u_t + u_{xxx} + uu_x + (Vu)_x + \gamma u = f, & t \in \mathbb{R},\ x \in \mathbb{T} \\ u(0, \cdot) = g(\cdot) \in H^s(\mathbb{T}). \end{cases} \tag{3.31}$$

Here $\gamma \geq 0$, and we assume that the potential V and the forcing f are mean-zero for each t, and they belong to $C_t^2 H_x^{s+\frac{3}{2}}(\mathbb{R} \times \mathbb{T})$. We also assume that the initial data g is mean-zero. In Chapter 5, we present various applications of the wellposedness methods developed here. In particular, we prove the existence and uniqueness of global attractors for equation (3.31).

Integrating the equation in x, we obtain

$$\frac{d}{dt}\int_0^{2\pi} u(t,x)dx + \gamma \int_0^{2\pi} u(t,x)dx = 0,$$

which implies that the mean is not conserved. However, note that in the case when the initial data is mean-zero, u remains mean-zero for all times. Similarly, using integration by parts, we have

$$\begin{aligned} \frac{d}{dt}\|u\|_{L_x^2}^2 &= -2\gamma\|u\|_2^2 + 2\int_0^{2\pi} fudx - \int_0^{2\pi} V_x u^2 dx \\ &\leq (-2\gamma + \|V_x\|_{L^\infty})\|u\|_{L^2}^2 + 2\|f\|_{L^2}\|u\|_{L^2} \\ &\leq (-2\gamma + M)\,\|u\|_{L^2}^2 + 2\|f\|_{L^2}\|u\|_{L^2}, \end{aligned}$$

where $M = \|V_x\|_{L_{t,x}^\infty}$. Setting

$$h(t) = e^{(2\gamma - M)t}\|u\|_{L^2}^2,$$

we obtain

$$h'(t) \leq 2e^{(\gamma - \frac{M}{2})t}\|f\|_{L^2}\sqrt{h(t)}.$$

This implies that

$$\frac{d}{dt}\sqrt{h(t)} \leq e^{(\gamma - \frac{M}{2})t}\|f\|_{L^2}. \tag{3.32}$$

Therefore

$$\|u(t)\|_{L^2} \leq e^{-(\gamma - \frac{M}{2})t}\|g\|_{L^2} + \int_0^t e^{-(\gamma - \frac{M}{2})(t-r)}\|f(r)\|_{L^2}dr.$$

In particular, when $V = 0$ we have

$$\|u(t)\|_{L^2} \leq e^{-\gamma t}\|g\|_{L^2} + \frac{\|f\|_{L_t^\infty L_x^2}}{\gamma}(1 - e^{-\gamma t}). \tag{3.33}$$

Using the estimates from the previous section, we will obtain the H^s local wellposedness of this equation for $s > -\frac{1}{2}$. Then the a priori L^2 bound above implies the L^2 global wellposedness.

Theorem 3.21 *Assume that $f, V \in C_t^2 H_x^{s+\frac{3}{2}}(\mathbb{R} \times \mathbb{T})$. Then the initial value problem (3.31) is locally wellposed for mean-zero data in H^s for $s > -\frac{1}{2}$. In particular, for*

$$\delta \lesssim (1 + \|g\|_{H^s})^{-6-}$$

with an implicit constant depending on γ, f, V, there is a unique solution $u \in Y_\delta^s \subset C_t^0 H_x^s([-\delta, \delta] \times \mathbb{T}))$ with

$$\|u\|_{X_\delta^{s,1/2}} \leq \|u\|_{Y_\delta^s} \leq C\|g\|_{H^s}.$$

Recall the following bounds from the previous section.

Lemma 3.22 *Let η be a smooth function supported on $[-2, 2]$, which is identically 1 on $[-1, 1]$. For any $\delta < 1$, the following a priori estimates hold for $s > -\frac{1}{2}$*

$$\|\eta(t)W_t g\|_{Y_\delta^s} \lesssim \|g\|_{H^s},$$

$$\left\|\eta(t)\int_0^t W_{t-s}F(s)ds\right\|_{Y_\delta^s} \lesssim \|F\|_{Z_\delta^s},$$

$$\|\partial_x(uv)\|_{Z_\delta^s} \lesssim \|u\|_{X_\delta^{s,1/2}}\|v\|_{X_\delta^{s,1/3}} + \|u\|_{X_\delta^{s,1/3}}\|v\|_{X_\delta^{s,1/2}}.$$

For any $-1/2 < b' < b < 1/2$

$$\|u\|_{X_\delta^{s,b'}} \lesssim \delta^{b-b'}\|u\|_{X_\delta^{s,b}}.$$

Finally

$$\|u\|_{Z_\delta^s} \lesssim \delta^{1-}\|u\|_{Y_\delta^s}.$$

The only statement in this lemma which is not already presented is the last one. By the fourth estimate of the lemma, it suffices to consider the second part of the Z_δ^s norm, which follows from

$$\left\|\frac{\langle k\rangle^s \mathcal{F}(u\eta(t/\delta))(\tau, k)}{\langle \tau - k^3\rangle}\right\|_{\ell_k^2 L_\tau^1}$$
$$\lesssim \left\|\langle k\rangle^s \mathcal{F}(u\eta(t/\delta))(\tau, k)\right\|_{\ell_k^2 L_\tau^{\infty-}} \left\|\frac{1}{\langle \tau - k^3\rangle}\right\|_{\ell_k^\infty L_\tau^{1+}}$$
$$\lesssim \|\delta\widehat{\eta}(\delta\tau)\|_{L_\tau^{\infty-}} \left\|\langle k\rangle^s \widehat{u}(\tau, k)\right\|_{\ell_k^2 L_\tau^1}$$
$$\lesssim \delta^{1-} \left\|\langle k\rangle^s \widehat{u}(\tau, k)\right\|_{\ell_k^2 L_\tau^1}.$$

The first inequality follows from Hölder, and the second one from Young's inequality in the τ variable.

To prove Theorem 3.21, we write

$$\Gamma u(t) = \eta(t)W_t g - \eta(t)\int_0^t W_{t-s}F(s)ds,$$

where

$$F = \frac{1}{2}\partial_x u^2 + (Vu)_x + \gamma u - f.$$

Applying the estimates in Lemma 3.22, one can easily see that

$$\begin{aligned}\|\Gamma u\|_{Y^s_\delta} &\lesssim \|g\|_{H^s} + \delta^{\frac{1}{6}-}\left(\|u\|^2_{Y^s_\delta} + \|u\|_{Y^s_\delta}\left(1 + \|V\|_{Y^s_\delta}\right) + \|f\|_{Y^s_\delta}\right)\\ &\lesssim \|g\|_{H^s} + \delta^{\frac{1}{6}-}\left(\|u\|^2_{Y^s_\delta} + \|u\|_{Y^s_\delta}\left(1 + \|V\|_{C^2_t H^{s+\frac{3}{2}}_x}\right) + \|f\|_{C^2_t H^{s+\frac{3}{2}}_x}\right),\end{aligned}$$

and similar estimates hold for the differences. Also noting that $\|V\|_{Y^s_\delta} \lesssim \|V\|_{C^2_t H^{s+\frac{3}{2}}_x}$ (see Exercise 3.4) , we see that Γ is a contraction on Y^s_δ, provided that

$$\delta \lesssim (1 + \|g\|_{H^s})^{-6-}, \tag{3.34}$$

with an implicit constant depending on γ, f, and V.

3.4 Differentiation by parts on the torus: unconditional wellposedness

We present below an alternative method initially developed by Babin–Ilyin–Titi [1] for proving local and global wellposedness of the KdV equation (3.23) for L^2 data on $\mathbb{T}$. As in the previous section, we restrict ourselves to mean-zero data. Precursors of the method in the case of initial value problems with periodic boundary conditions were developed for the Euler and Navier–Stokes equations by Babin–Mahalov–Nicolaenko in [2], and for water wave problems by Embid–Majda [54] and by Germain–Masmudi–Shatah [64].

The main idea of the method is a normal form transformation in the spirit of Shatah [127]. The idea is best illustrated by a special case of averaging in ordinary differential equations. Let Ω be large and $x \in \mathbb{R}^n$, and consider the equation

$$\frac{dx}{dt} = e^{i\Omega t}f(x).$$

We can rewrite the right-hand side as

$$\frac{dx}{dt} = \frac{d}{dt}\left(\frac{e^{i\Omega t}}{i\Omega}f(x)\right) - \frac{e^{i\Omega t}}{i\Omega}\frac{d}{dt}f(x)$$

$$= \frac{d}{dt}\left(\frac{e^{i\Omega t}}{i\Omega}f(x)\right) - \frac{e^{i\Omega t}}{i\Omega}f'(x)e^{i\Omega t}f(x).$$

Therefore

$$\frac{d}{dt}\left(x - \frac{e^{i\Omega t}}{i\Omega}f(x)\right) = -\frac{e^{i\Omega t}}{i\Omega}f'(x)e^{i\Omega t}f(x).$$

Assuming that f and its derivative are uniformly bounded, and integrating the last relation, one obtains the standard bound

$$\|x(t) - x(0)\| = O(\Omega^{-1})$$

for $t = O(1)$.

In the case of the KdV equation, we will have similar calculations on the Fourier side after a change of variable similar to the one in the definition (3.16) of $X^{s,b}$ spaces. The Ω gain in the previous argument corresponds to a gain in the Fourier variable, eliminating the derivative in the nonlinearity. One has to be careful with the resonant (nonoscillating) terms, and apply the differentiation by parts twice for this method to work. Moreover, to close the contraction one has to consider high and low frequencies separately.

We use the Fourier series representation

$$u(t, x) = \sum_{k\neq 0} u_k(t)e^{ikx}$$

with

$$u_k := \widehat{u}(k) = \frac{1}{2\pi}\int_0^{2\pi} u(t, x)e^{-ikx}dx.$$

Since we work with mean-zero solutions, there is no $k = 0$ term above. We also note that $u_k = \overline{u_{-k}}$ since u is real valued. We write the KdV equation

$$u_t + u_{xxx} + uu_x = 0,$$

on the Fourier side as

$$\partial_t u_k = -\frac{ik}{2}\sum_{k_1+k_2=k} u_{k_1}u_{k_2} + ik^3u_k.$$

Then using the identity

$$(k_1 + k_2)^3 - k_1^3 - k_2^3 = 3(k_1 + k_2)k_1k_2,$$

and the transformation

$$u_k(t) = v_k(t)e^{ik^3 t}$$

the equation can be written in the form

$$\partial_t v_k = -\frac{ik}{2} \sum_{k_1+k_2=k} e^{-i3kk_1k_2t} v_{k_1} v_{k_2}. \tag{3.35}$$

The variables v_k are often called interaction variables, and their time behavior is smoother than the original variables. Note the similarity of this transformation with the definition (3.16) of $X^{s,b}$ spaces.

Below, we perform the differentiation by parts process in the t variable. The justification of the steps is presented at the end. We define

$$B_2(f, g)_0 = \rho_0 = D_3(f, g, h)_0 = B_3(f, g, h)_0 = 0,$$

and for $k \neq 0$, we define

$$B_2(f, g)_k = -\frac{1}{6} \sum_{k_1+k_2=k} \frac{e^{-3ikk_1k_2t} f_{k_1} g_{k_2}}{k_1 k_2},$$

$$\rho_k = -\frac{i}{12k} v_k |v_k|^2,$$

$$D_3(f, g, h)_k = \frac{i}{12} \sum_{\substack{k_1+k_2+k_3=k \\ (k_1+k_2)(k_1+k_3)(k_2+k_3)\neq 0}} \frac{e^{-3it(k_1+k_2)(k_2+k_3)(k_3+k_1)}}{k_1} f_{k_1} g_{k_2} h_{k_3},$$

$$B_3(f, g, h)_k = -\frac{1}{36} \sum_{\substack{k_1+k_2+k_3=k \\ (k_1+k_2)(k_1+k_3)(k_2+k_3)\neq 0}} \frac{e^{-3it(k_1+k_2)(k_2+k_3)(k_3+k_1)}}{k_1(k_1+k_2)(k_2+k_3)(k_3+k_1)} f_{k_1} g_{k_2} h_{k_3}.$$

Using the identity

$$e^{-i3kk_1k_2t} = -\partial_t\Big(\frac{1}{3ikk_1k_2} e^{-i3kk_1k_2t}\Big),$$

and the symmetry of B_2, we can rewrite (3.35) as follows

$$\partial_t v_k = -\partial_t B_2(v, v)_k + 2B_2(v, v_t)_k.$$

Using (3.35), we have

$$B_2(v, v_t)_k = -\frac{1}{6} \sum_{k_1+k_2=k} \frac{e^{-3ikk_1k_2t}}{k_1k_2} v_{k_1} \partial_t v_{k_2}$$
$$= \frac{i}{12} \sum_{k=k_1+k_2} \frac{e^{-3ikk_1k_2t}}{k_1} v_{k_1} \Big(\sum_{\mu+\lambda=k_2} e^{-3itk_2\mu\lambda} v_\mu v_\lambda \Big)$$
$$= \frac{i}{12} \sum_{k=k_1+\mu+\lambda} \frac{v_{k_1} v_\mu v_\lambda}{k_1} e^{-3it[kk_1(\mu+\lambda)+\mu\lambda(\mu+\lambda)]}.$$

We note that $\mu + \lambda$ cannot be zero since $\mu + \lambda = k_2$ and $v_0 = 0$. Using the identity

$$kk_1 + \mu\lambda = (k_1 + \mu + \lambda)k_1 + \mu\lambda = (k_1 + \mu)(k_1 + \lambda)$$

and by renaming the variables $k_2 = \mu, k_3 = \lambda$, we have that

$$B_2(v, v_t)_k = \frac{i}{12} \sum_{\substack{k_1+k_2+k_3=k \\ k_2+k_3\neq 0}} \frac{e^{-3it(k_1+k_2)(k_2+k_3)(k_3+k_1)}}{k_1} v_{k_1} v_{k_2} v_{k_3}.$$

Combining the formulas above, we have that

$$\partial_t (v_k + B_2(v, v)_k) = D_3(v, v, v)_k + D_3^r(v, v, v)_k,$$

where D_3^r is defined as

$$D_3^r(f, g, h)_k = \frac{i}{12} \sum_{\substack{k_1+k_2+k_3=k \\ k_2+k_3\neq 0,\ (k_1+k_2)(k_1+k_3)=0}} \frac{f_{k_1} g_{k_2} h_{k_3}}{k_1}.$$

Here the superscript r stands for the resonant terms

$$(k_1 + k_2)(k_3 + k_1) = 0, \quad k_2 + k_3 \neq 0. \tag{3.36}$$

The set for which (3.36) holds is the disjoint union of the following three sets

$$R_1 = \{k_1 + k_2 = 0\} \cap \{k_3 + k_1 = 0\} \Leftrightarrow \{k_1 = -k,\ k_2 = k,\ k_3 = k\},$$

$$R_2 = \{k_1 + k_2 = 0\} \cap \{k_3 + k_1 \neq 0\} \Leftrightarrow \{k_1 = j,\ k_2 = -j,\ k_3 = k,\ |j| \neq |k|\},$$

$$R_3 = \{k_3 + k_1 = 0\} \cap \{k_1 + k_2 \neq 0\}\} \Leftrightarrow \{k_1 = j,\ k_2 = k,\ k_3 = -j,\ |j| \neq |k|\}.$$

Thus

$$D_3^r(v,v,v)_k = \frac{i}{12}\sum_{\lambda=1}^{3}\sum_{R_\lambda}\frac{v_{k_1}v_{k_2}v_{k_3}}{k_1}$$
$$= \frac{i}{12}\frac{v_{-k}v_kv_k}{-k} + \frac{i}{12}v_k\sum_{|j|\neq 0,|k|}\frac{v_jv_{-j}}{j} + \frac{i}{12}v_k\sum_{|j|\neq 0,|k|}\frac{v_jv_{-j}}{j}.$$

Note that the second and third terms in the sum above are identically zero due to the symmetry relation $j \leftrightarrow -j$. Thus

$$D_3^r(v,v,v)_k = -\frac{i}{12k}v_k|v_k|^2 = \rho_k,$$

where we used $v_{-k} = \overline{v}_k$. This implies that

$$\partial_t\left(v_k + B_2(v,v)_k\right) = \rho_k + D_3(v,v,v)_k. \tag{3.37}$$

Since the exponent in the last term is not zero, we can differentiate by parts one more time using

$$D_3(v,v,v)_k = \partial_t B_3(v,v,v)_k - B_3(v_t,v,v)_k - B_3(v,v_t,v)_k - B_3(v,v,v_t)_k$$
$$= \partial_t B_3(v,v,v)_k - B_3(v_t,v,v)_k - 2B_3(v,v_t,v)_k.$$

As before, we express time derivatives using (3.35). From now on, $\sum^*$ means that the sum is over all indices for which the denominator does not vanish. A calculation reveals that

$$B_3(v_t,v,v)_k = \frac{i}{72}\sum_{k_1+k_2+k_3+k_4=k}^{*}\frac{e^{it\psi(k_1,k_2,k_3,k_4)}}{(k_1+k_2)(k_1+k_3+k_4)(k_2+k_3+k_4)}v_{k_1}v_{k_2}v_{k_3}v_{k_4}$$

and

$$B_3(v,v_t,v)_k = \frac{i}{72}\sum_{k_1+k_2+k_3+k_4=k}^{*}\frac{e^{it\psi(k_1,k_2,k_3,k_4)}(k_3+k_4)}{k_1(k_1+k_2)(k_1+k_3+k_4)(k_2+k_3+k_4)}v_{k_1}v_{k_2}v_{k_3}v_{k_4}.$$

The phase function ψ will be irrelevant for our calculations since it is going to be estimated out by taking absolute values inside the sums.

If we put everything together and denote

$$B_4(v) = -B_3(v_t,v,v) - 2B_3(v,v_t,v),$$

we obtain

$$\partial_t(v_k - B(v)_k) = \rho_k + B_4(v)_k, \tag{3.38}$$

where

$$B(v)_k = -B_2(v,v)_k + B_3(v,v,v)_k,$$

$$B_4(v)_k = -\frac{i}{72} \sum_{k_1+k_2+k_3+k_4=k}^{*} \frac{e^{i\psi(k_1,k_2,k_3,k_4)t}(2k_3 + 2k_4 + k_1)v_{k_1}v_{k_2}v_{k_3}v_{k_4}}{k_1(k_1+k_2)(k_1+k_3+k_4)(k_2+k_3+k_4)}.$$

Integrating (3.38) from 0 to t, we obtain

$$v_k(t) = v_k(0) + B(v)_k(t) - B(v)_k(0) + \int_0^t (\rho_k + B_4(v)_k)(s)ds. \tag{3.39}$$

Note that if we integrate the original equation (3.35), we obtain

$$v_k(t) = v_k(0) - \frac{ik}{2} \int_0^t \sum_{k_1+k_2=k} e^{-i3kk_1k_2s} v_{k_1}v_{k_2}ds. \tag{3.40}$$

Fix N large to be determined later. Define the operator Γ_N as follows

$$\Gamma_N(v)_k(t) = \begin{cases} v_k(0) + B(v)_k(t) - B(v)_k(0) + \int_0^t (\rho_k + B_4(v)_k)(s)ds, & |k| > N \\ v_k(0) - \frac{ik}{2} \int_0^t \sum_{k_1+k_2=k} e^{-i3kk_1k_2s} v_{k_1}v_{k_2}ds, & |k| \le N \end{cases} \tag{3.41}$$

The following proposition includes all the a priori estimates we need to prove that Γ_N is a contraction on a suitable space.

Proposition 3.23 *We have the following a priori bounds*

$$\|B(v)\|_{\ell^2_{|k|>N}} \lesssim \frac{1}{N^{1/4}} \left(\|v\|_{\ell^2}^2 + \|v\|_{\ell^2}^3 \right), \tag{3.42}$$

$$\|B_4(v)\|_{\ell^2_{|k|>N}} \lesssim \|v\|_{\ell^2}^4, \tag{3.43}$$

$$\|\rho_k\|_{\ell^2_{|k|>N}} \lesssim \frac{1}{N} \|v\|_{\ell^2}^3, \tag{3.44}$$

$$\left\| k \sum_{k_1+k_2=k} e^{i3kk_1k_2s} v_{k_1}v_{k_2} \right\|_{\ell^2_{|k|\le N}} \lesssim N^{3/2} \|v\|_{\ell^2}^2. \tag{3.45}$$

Moreover, each operator above is Lipschitz continuous from ℓ^2 to ℓ^2.

Using this proposition, we now prove that Γ_N is a contraction on

$$X = \left\{ v \in C_t^0 \ell_k^2([-\delta,\delta] \times (\mathbb{Z}\setminus\{0\})) : \|v\|_{L^\infty_{[-\delta,\delta]}\ell^2} \le 2\|g\|_{L^2} \right\},$$

where N, δ depends on $\|g\|_{L^2}$. Indeed

$$\|\Gamma_N(v)\|_{L^\infty_{[-\delta,\delta]}\ell^2} \le \|v(0)\|_{L^2} + C\frac{1}{N^{1/4}} \left(\|v\|^2_{L^\infty_{[-\delta,\delta]}\ell^2} + \|v\|^3_{L^\infty_{[-\delta,\delta]}\ell^2} \right)$$
$$+ C\delta \frac{1}{N} \|v\|^3_{L^\infty_{[-\delta,\delta]}\ell^2} + C\delta \|v\|^4_{L^\infty_{[-\delta,\delta]}\ell^2} + C\delta N^{3/2} \|v\|^2_{L^\infty_{[-\delta,\delta]}\ell^2}.$$

Similar estimates hold for the difference $\Gamma_N v_1 - \Gamma_N v_2$ by Proposition 3.23. The fact that

$$\Gamma_N(v) \in C_t^0 \ell_k^2([-\delta,\delta] \times (\mathbb{Z}\backslash\{0\}))$$

also follows from the Lipschitz continuity statement in Proposition 3.23.

Choosing first N large depending on $\|g\|_{L^2}$, and then δ small depending on N and $\|g\|_{L^2}$, we see that Γ_N is a contraction on X. This gives us a unique solution in X and continuous dependence on initial data for the equation $\Gamma_N(v) = v$.

We now define u to be a solution of the KdV equation on $[-\delta,\delta]$ if $u \in C_t^0 L_x^2([-\delta,\delta] \times \mathbb{T})$, and if for each $k \in \mathbb{Z}$, $t \in [-\delta,\delta]$, $v_k(t) = u_k(t)e^{-ik^3 t}$ satisfies (3.40). Note that this agrees with the notion of the solution in Definition 3.1 with $X_\delta = C_t^0 L_x^2([-\delta,\delta] \times \mathbb{T})$.

Solutions defined as above also satisfy (3.35) for each k and t. Moreover, we have the bound

$$\sup_{t\in[-\delta,\delta]} |\partial_t v_k| \lesssim |k| \tag{3.46}$$

with the implicit constant depending only on $\|v\|_{L^\infty_{[-\delta,\delta]}\ell^2}$. For any given such solution of the KdV equation, the bound (3.46) suffices to justify the differentiation by parts process. Indeed, it suffices to check that for each k one can change the order of summation and differentiation, which follows from (3.46), the L^2 bound on v, and the mean value theorem. Therefore, any given solution is also a fixed point of the operator Γ_N above for each fixed N. In particular, the smooth solutions of KdV, which exists by the energy method presented in Section 3.1, are also fixed points of Γ_N for each N.

We now prove that given L^2 initial data, g, the fixed point v of Γ_N, also satisfies (3.40) for each k. To do this, we approximate g by a smooth sequence, g^n, and obtain the corresponding solutions v^n of the KdV equation. Since v^n are also fixed points of Γ_N, by continuous dependence on initial data v^n converges to v in $C_t^0 \ell_k^2$. Therefore, for each fixed k, taking the limit as $n \to \infty$ in (3.40), we see that v also satisfies (3.40). This also proves that the fixed points of Γ_N satisfy the L^2 conservation law.

Finally, we prove the uniqueness of the solution of KdV. Let

$$v^1, v^2 \in C_t^0 \ell_k^2 ([-\delta,\delta] \times (\mathbb{Z}\backslash\{0\}))$$

be two solutions of KdV with the same initial data in L^2. Let

$$t_0 = \sup\left\{t \geq 0 : v^1(s) = v^2(s), s \in [0,t]\right\}.$$

If $t_0 = \delta$, there is nothing to prove. Assume on the contrary that $t_0 < \delta$. By continuity, we can find $\delta_1 > 0$ so that

$$\left\|v^1(t)\right\|_{\ell^2}, \left\|v^2(t)\right\|_{\ell^2} \leq 2\left\|v^1(t_0)\right\|_{\ell^2} \quad \text{for } t \in [t_0, t_0 + \delta_1].$$

By the remark above, v^1 and v^2 are fixed points of Γ_N in $X_{[t_0,t_0+\delta_1]}$. Therefore, $v^1 = v^2$ on $[t_0, t_0+\delta_1]$, which is a contradiction.

Remark 3.24 Uniqueness as it is proved above is known as "unconditional uniqueness" in the literature. We remind the reader that the methods used in the previous two sections give uniqueness only in a proper subset of $C_t^0 H_x^s$. Also see Kwon–Oh [100] for the unconditional uniqueness of the modified KdV equation on the torus, and see Shung–Guo–Kwon–Oh [40] for the unconditional uniqueness of a quadratic derivative Schrödinger equation on the torus. Both papers used a variation of the method in Babin–Ilyin–Titi [1].

It remains to prove Proposition 3.23.

Proof of Proposition 3.23 We start with (3.42). For $|k| > N$, we have

$$|B(v)_k| \lesssim \frac{1}{N} \sum_{k_1+k_2=k} \frac{|v_{k_1}||v_{k_2}|}{|k_2|} + \frac{1}{N^{\frac{1}{4}}} \sum^{*}_{k_1+k_2+k_3=k} \frac{|v_{k_1}||v_{k_2}||v_{k_3}|}{|k_1||k_2|^{3/4}}.$$

The first one follows assuming by symmetry that $|k_1| \gtrsim |k|$, while the second follows using

$$|(k_1+k_2)(k_1+k_3)(k_2+k_3)| \gtrsim \max(|k_1|, |k_2|, |k_3|) \gtrsim |k_2|^{\frac{3}{4}} |k|^{\frac{1}{4}},$$

when the left-hand side is not zero. Taking the ℓ^2 norm, we have

$$\begin{aligned} \|B(v)\|_{\ell^2_{|k|>N}} &\lesssim \frac{1}{N} \Big\| |v_k| * \frac{|v_k|}{|k|} \Big\|_{\ell^2} + \frac{1}{N^{1/4}} \Big\| \frac{|v_k|}{|k|} * \frac{|v_k|}{|k|^{3/4}} * |v_k| \Big\|_{\ell^2} \\ &\lesssim \frac{1}{N} \|v\|_{\ell^2} \Big\| \frac{v_k}{k} \Big\|_{\ell^1} + \frac{1}{N^{1/4}} \|v\|_{\ell^2} \Big\| \frac{v_k}{k} \Big\|_{\ell^1} \Big\| \frac{v_k}{|k|^{3/4}} \Big\|_{\ell^1} \\ &\lesssim \frac{1}{N^{1/4}} (\|v\|_{\ell^2}^2 + \|v\|_{\ell^2}^3), \end{aligned}$$

where we used Young's inequality and the Cauchy–Schwarz inequality.

The inequality (3.44) follows immediately from $\ell^2 \subset \ell^\infty$. To prove (3.45), we note

$$\Big\| k \sum_{k_1+k_2=k} e^{i3kk_1k_2s} v_{k_1} v_{k_2} \Big\|_{\ell^2_{|k|\leq N}} \lesssim \|v * v\|_{\ell^\infty} \|k\|_{\ell^2_{|k|\leq N}} \lesssim N^{3/2} \|v\|_{\ell^2}^2.$$

It remains to prove (3.43). We estimate B_4 as

$$\begin{aligned} |B_4(v)_k| \lesssim & \sum^{*}_{k_1+k_2+k_3+k_4=k} \frac{|v_{k_1} v_{k_2} v_{k_3} v_{k_4}|}{|k_1+k_2||k_1+k_3+k_4||k_2+k_3+k_4|} \\ & + \sum^{*}_{k_1+k_2+k_3+k_4=k} \frac{|v_{k_1} v_{k_2} v_{k_3} v_{k_4}|}{|k_1||k_1+k_2||k_2+k_3+k_4|}. \end{aligned}$$

We will estimate the first line; the same method works for the second one. By duality, it suffices to estimate

$$\sup_{\|h\|_{\ell^2}=1} \sum^*_{k_1,k_2,k_3,k_4} \frac{|v_{k_1} v_{k_2} v_{k_3} v_{k_4}||h_{k_1+k_2+k_3+k_4}|}{|k_1+k_2||k_1+k_3+k_4||k_2+k_3+k_4|}$$

$$\leq \sup_{\|h\|_{\ell^2}=1} \left[\sum^*_{k_1,k_2,k_3,k_4} \frac{|v_{k_1} v_{k_4}|^2}{|k_1+k_2|^2 |k_1+k_3+k_4|^2} \right]^{1/2} \left[\sum^*_{k_1,k_2,k_3,k_4} \frac{|v_{k_2} v_{k_3}|^2 |h_{k_1+k_2+k_3+k_4}|^2}{|k_2+k_3+k_4|^2} \right]^{1/2}$$

$$\lesssim \|v\|^4_{\ell^2}.$$

The estimate for the first sum follows by summing in the order k_2, k_3, k_1, k_4. For the second, we sum in the order k_1, k_4, k_2, k_3.

Finally, the Lipschitz continuity follows from the bounds above by considering the differences as usual. □

3.5 Local theory for NLS on the torus

In this section, we study the NLS equation on the torus

$$\begin{cases} iu_t + u_{xx} \pm |u|^{p-1}u = 0, & x \in \mathbb{T}, \quad t \in \mathbb{R}, \\ u(0,\cdot) = g(\cdot) \in H^s(\mathbb{T}), \end{cases} \tag{3.47}$$

for $s \geq 0$. We restrict our discussion to the cubic $p = 3$ and the quintic $p = 5$ cases, which are physically the most relevant cases. It is easy to see that the smooth solutions to the equation satisfy the L^2 norm conservation. Also note that the energy

$$H(u) = \frac{1}{2}\|u_x\|^2_{L^2(\mathbb{T})} \mp \frac{1}{p+1}\|u\|^{p+1}_{L^{p+1}(\mathbb{T})}$$

is conserved.

The local wellposedness in H^s, $s > \frac{1}{2}$, of nonlinear Schrödinger equations with power type nonlinearities on the real line or the torus, follows immediately from the algebra property of Sobolev spaces; see Exercise 3.5. Recall that this is not the case for the KdV equation because of the derivative nonlinearity. On the real line, for $p < 5$, Strichartz estimates (see Section 2.1) suffice to extend the local wellposedness theory to $H^s(\mathbb{R})$, $s \geq 0$; see Tsutsumi [151]. Because of the L^2 norm conservation, the solution is actually global in $L^2(\mathbb{R})$.

The quintic NLS on the real line is also locally wellposed in $L^2(\mathbb{R})$, see Cazenave–Weissler [31]. In this case, the local existence time depends on the profile of the initial data, and hence the L^2 norm conservation is not enough to obtain global wellposedness. In fact, in the focusing case (the minus sign in front of the nonlinearity) the solution may blow up in finite time even for H^1 data; see Glassey [68], and Ogawa–Tsutsumi [118]. The question of global wellposedness in $L^2(\mathbb{R})$ of the defocusing quintic NLS was open for a long time, and it was answered recently in the affirmative by Dodson [52]. In a similar fashion, Dodson [53] obtained the L^2 global wellposedness of the focusing quintic NLS when the L^2 norm on the initial data is less than the L^2 norm of the ground state. This result is sharp since the focusing problem has blowup solutions with data having the L^2 norm of the ground state; see Exercise 3.14.

For the problem on the torus, the Strichartz estimates (Theorem 2.10) are not sufficient to obtain the L^2 local wellposedness of (3.47) since there is a derivative loss in the L^6 estimate. However, using the $X^{s,b}$ space methods we developed in the previous sections together with the Strichartz estimates we can establish the low regularity local wellposedness of the cubic and the quintic NLS on the torus, which was obtained by Bourgain in [15].

3.5.1 L^2 wellposedness of cubic NLS on the torus

We now establish the local theory of cubic NLS

$$\begin{cases} iu_t + u_{xx} \pm |u|^2 u = 0, & x \in \mathbb{T}, \quad t \in \mathbb{R}, \\ u(0,\cdot) = g(\cdot) \in H^s(\mathbb{T}) \end{cases} \tag{3.48}$$

for $s \geq 0$ using $X^{s,b}$ spaces. We will see that the range of s is optimal in the next section. The corresponding $X^{s,b}$ space is defined via the norm

$$\|u\|_{X^{s,b}} = \|e^{-it\partial_{xx}} u\|_{H^s_x H^b_t} = \left\| \langle k \rangle^s \langle \tau + k^2 \rangle^b \widehat{u}(\tau, k) \right\|_{L^2_\tau \ell^2_k}. \tag{3.49}$$

The restricted norm $X^{s,b}_\delta$ is defined analogously, and the lemmas 3.9, 3.10, 3.11, and 3.12 we proved in Section 3.3 remain valid. We also have the following variant of Theorem 3.18; see Bourgain [15]. As in the case of the KdV equation, this refinement allows us to extract a power of δ in the wellposedness proof. We leave the proof of this theorem to the reader in Exercise 3.9.

Theorem 3.25 *For any smooth space-time function u, we have*

$$\|u\|_{L^4_{x\in\mathbb{T},\, t\in\mathbb{R}}} \lesssim \|u\|_{X^{0,3/8}}.$$

Proposition 3.26 *For any $s \geq 0$, we have*

$$\left\| |u|^2 u \right\|_{X^{s,-\frac{3}{8}}_\delta} \lesssim \|u\|^2_{X^{0,\frac{3}{8}}_\delta} \|u\|_{X^{s,\frac{3}{8}}_\delta}.$$

Proof By Plancherel and L^2 duality, we have

$$\left\||u|^2u\right\|_{X^{s,-\frac{3}{8}}} = \sup_{\|v\|_{X^{-s,\frac{3}{8}}}=1} \left|\langle |u|^2u, v\rangle_{L^2(\mathbb{T}\times\mathbb{R})}\right|.$$

Note that

$$\begin{aligned}\left|\langle |u|^2u, v\rangle_{L^2(\mathbb{T}\times\mathbb{R})}\right| &= \left|\left\langle J^s(|u|^2u), J^{-s}v\right\rangle_{L^2(\mathbb{T}\times\mathbb{R})}\right| \\ &\leq \left\|J^s(|u|^2u)\right\|_{L^{\frac{4}{3}}(\mathbb{T}\times\mathbb{R})} \left\|J^{-s}v\right\|_{L^4(\mathbb{T}\times\mathbb{R})} \\ &\leq \|J^s u\|_{L^4(\mathbb{T}\times\mathbb{R})}\|u\|^2_{L^4(\mathbb{T}\times\mathbb{R})}\|v\|_{X^{-s,\frac{3}{8}}(\mathbb{T}\times\mathbb{R})} \\ &\lesssim \|u\|^2_{X^{0,\frac{3}{8}}}\|u\|_{X^{s,\frac{3}{8}}}.\end{aligned}$$

In the second inequality, we used Lemma 1.11, while we used Theorem 3.25 in the second and third inequalities. □

Theorem 3.27 *For any $s \geq 0$ with $g \in H^s(\mathbb{T})$, the equation* (3.48) *is locally wellposed in $X^{s,b}_\delta$ for any $\frac{1}{2} < b < \frac{5}{8}$.*

Proof We have that

$$\Gamma u(t) = e^{it\partial_{xx}}g \pm i\int_0^t e^{i(t-s)\partial_{xx}}|u|^2u(s)ds.$$

Using lemmas 3.10, 3.11, and 3.12, and Proposition 3.26, we estimate

$$\begin{aligned}\|\Gamma u(t)\|_{X^{s,b}_\delta} &\lesssim \|g\|_{H^s(\mathbb{T})} + \delta^{1-b-\frac{3}{8}}\left\||u|^2u\right\|_{X^{s,-\frac{3}{8}}_\delta} \\ &\lesssim \|g\|_{H^s(\mathbb{T})} + \delta^{1-b-\frac{3}{8}}\|u\|^2_{X^{0,\frac{3}{8}}_\delta}\|u\|_{X^{s,\frac{3}{8}}_\delta} \\ &\lesssim \|g\|_{H^s(\mathbb{T})} + \delta^{1-b-}\|u\|^2_{X^{0,b}_\delta}\|u\|_{X^{s,b}_\delta} \\ &\lesssim \|g\|_{H^s(\mathbb{T})} + \delta^{1-b-}\|u\|^3_{X^{s,b}_\delta}.\end{aligned}$$

We omit the standard details. We note that since $b > \frac{1}{2}$ by lemma 3.9, the solution is in fact a continuous function with values in $H^s(\mathbb{T})$. □

Theorem 3.28 *Equation* (3.48) *is globally wellposed in $H^s(\mathbb{T})$ for any $s \geq 0$.*

Proof For $s = 0$, the statement follows easily since the L^2 norm of the solution is conserved. The case when $s \geq 1$ is outlined in the exercises. The case $0 < s < 1$ requires some care and we include it here for completeness. We first observe by the local theory in L^2 that we can find δ_0 depending on $\|g\|_{L^2}$ so that

$$\|u\|_{X^{0,b}_{\delta_0}} \lesssim \|g\|_{L^2(\mathbb{T})}, \quad \frac{1}{2} < b < \frac{5}{8}.$$

Note that for any smooth solution u, the bound

$$\|u(t)\|_{X_\delta^{s,b}} \lesssim \|g\|_{H^s(\mathbb{T})} + \delta^{1-b-} \|u\|_{X_\delta^{0,b}}^2 \|u\|_{X_\delta^{s,b}}$$

holds for any $\delta > 0$ and $\frac{1}{2} < b < \frac{5}{8}$ (see the proof of Theorem 3.27). Taking $\delta < \delta_0$, we obtain

$$\|u(t)\|_{X_\delta^{s,b}} \lesssim \|g\|_{H^s(\mathbb{T})} + \delta^{1-b-} \|g\|_{L^2}^2 \|u\|_{X_\delta^{s,b}}.$$

Therefore, for some $\delta_1 \leq \delta_0$ depending only on $\|g\|_{L^2}$, we have

$$\|u\|_{X_{\delta_1}^{s,b}} \lesssim \|g\|_{H^s(\mathbb{T})}.$$

This estimate and the fact that $b > \frac{1}{2}$ implies the a priori bound

$$\|u(t)\|_{H^s(\mathbb{T})} \leq C \|g\|_{H^s(\mathbb{T})}$$

for $t \in [0, \delta_1]$. Since δ_1 depends only on the L^2 norm, we can iterate this inequality and obtain

$$\|u(t)\|_{H^s(\mathbb{T})} \leq C^{|t|} \|g\|_{H^s(\mathbb{T})}.$$

This a priori bound extends to any H^s solution by smooth approximation in a standard way. This finishes the proof of global wellposedness. □

3.5.2 H^s local wellposedness of the quintic NLS on the torus

In this section, we established the local wellposedness of the quintic NLS on the torus

$$\begin{cases} iu_t + u_{xx} \pm |u|^4 u = 0, & x \in \mathbb{T}, \quad t \in \mathbb{R}, \\ u(0, \cdot) = g(\cdot), \end{cases} \tag{3.50}$$

in $H^s(\mathbb{T})$, $s > 0$. The wellposedness theory for data in $L^2(\mathbb{T})$ is an open problem. It is known that the L^2 theory cannot be established using Banach fixed point arguments; see Kishimoto [95].

The local wellposedness is proved by a contraction argument in $X^{s,b}$ spaces using the L^6 Strichartz estimate, Theorem 2.10. Since there is a derivative loss in this theorem, one needs to be careful with the nonlinear interaction of different frequency components. The proof we present is somehow different than the proof given by Bourgain in [15]. It is based on an argument that uses multilinear Strichartz estimates; see, e.g., Burq–Gérard–Tzvetkov [24], Catoire–Wang [27], and Demirbaş [50].

We start with a trilinear refinement of Theorem 2.10. First note that since the L^6 norm is modulation invariant for any interval I, Theorem 2.10 implies that

$$\left\|Q_I e^{it\partial_{xx}} g\right\|_{L^6_{x,t\in\mathbb{T}}} \lesssim |I|^{\epsilon} \|g\|_{L^2}, \tag{3.51}$$

for any $\epsilon > 0$, where $|I|$ is the length of the interval I, and Q_I is the projection onto the frequencies in the interval I; see Exercise 3.15. Moreover, as in Corollary 3.17, the inequality (3.51) implies that

$$\|Q_I u\|_{L^6_{x\in\mathbb{T}} L^6_{t\in[-\delta,\delta]}} \lesssim |I|^{\epsilon} \|u\|_{X^{0,b}_{\delta}} \tag{3.52}$$

for any $\epsilon > 0$, $b > \frac{1}{2}$. As usual, we ignore the δ dependence till the end of the proof.

Lemma 3.29 *For any dyadic $N_1 \le N_2 \le N_3$, we have*

$$\left\|P_{N_1} u P_{N_2} v P_{N_3} w\right\|_{L^2_{x,t\in\mathbb{T}}} \lesssim N_2^{\epsilon} \|P_{N_1} u\|_{X^{0,b}} \|P_{N_2} v\|_{X^{0,b}} \|P_{N_3} w\|_{X^{0,b}}, \tag{3.53}$$

for any $\epsilon > 0$ and $b > \frac{1}{2}$.

Proof We write

$$\begin{aligned} P_{N_1} u P_{N_2} v P_I P_{N_3} w &= \sum_I P_{N_1} u P_{N_2} v Q_I P_{N_3} w \\ &= \sum_I P_{20I} \left(P_{N_1} u P_{N_2} v Q_I P_{N_3} w\right), \end{aligned}$$

where the sum is over disjoint intervals of length N_2, and $20I$ is the interval with the same center as I and 20 times the length. In the second equality, we used the fact that $\mathcal{F}(P_{N_1} u P_{N_2} v Q_I P_{N_3} w)$ is supported in $20I$. Therefore, by almost orthogonality, we have

$$\left\|P_{N_1} u P_{N_2} v P_{N_3} w\right\|^2_{L^2_{x,t\in\mathbb{T}}} \lesssim \sum_I \left\|P_{N_1} u P_{N_2} v Q_I P_{N_3} w\right\|^2_{L^2_{x,t\in\mathbb{T}}}.$$

By Hölder's inequality, (3.52), and $Q_I = Q_I^2$, we estimate this by

$$\begin{aligned} \sum_I \left\|P_{N_1} u P_{N_2} v Q_I P_{N_3} w\right\|^2_{L^2_{x,t\in\mathbb{T}}} &\le \left\|P_{N_1} u\right\|^2_{L^6_{x,t\in\mathbb{T}}} \left\|P_{N_2} v\right\|^2_{L^6_{x,t\in\mathbb{T}}} \sum_I \left\|Q_I P_{N_3} w\right\|^2_{L^6_{x,t\in\mathbb{T}}} \\ &\lesssim N_2^{\epsilon} \left\|P_{N_1} u\right\|^2_{X^{0,b}} \left\|P_{N_2} v\right\|^2_{X^{0,b}} \sum_I \left\|Q_I P_{N_3} w\right\|^2_{X^{0,b}} \\ &\lesssim N_2^{\epsilon} \left\|P_{N_1} u\right\|^2_{X^{0,b}} \left\|P_{N_2} v\right\|^2_{X^{0,b}} \left\|P_{N_3} w\right\|^2_{X^{0,b}}. \end{aligned}$$

This finishes the proof of the lemma. □

Corollary 3.30 *Given $\epsilon > 0$, there exists $b' < \frac{1}{2}$ such that for any dyadic $N_1 \le N_2 \le N_3$ we have*

$$\left\| P_{N_1} u P_{N_2} v P_{N_3} w \right\|_{L^2_{x,t\in\mathbb{T}}} \lesssim N_2^{\epsilon} \|P_{N_1} u\|_{X^{0,b'}} \|P_{N_2} v\|_{X^{0,b'}} \|P_{N_3} w\|_{X^{0,b'}}. \tag{3.54}$$

Proof First note that by interpolating the inequalities

$$\|u\|_{L^\infty_t L^2_x} \lesssim \|u\|_{X^{0,\frac{1}{2}+}}, \quad \|u\|_{L^2_t L^2_x} = \|u\|_{X^{0,0}},$$

we obtain

$$\|u\|_{L^6_t L^2_x} \lesssim \|u\|_{X^{0,\frac{1}{3}+}}.$$

Therefore, by Hölder's inequality and Sobolev embedding

$$\begin{aligned}
\left\| P_{N_1} u P_{N_2} v P_{N_3} w \right\|_{L^2_{x,t\in\mathbb{T}}} &\lesssim \|P_{N_1} u\|_{L^6_t L^\infty_x} \|P_{N_2} v\|_{L^6_t L^\infty_x} \|P_{N_3} w\|_{L^6_t L^2_x} \\
&\lesssim N_2^{1+} \|P_{N_1} u\|_{L^6_t L^2_x} \|P_{N_2} v\|_{L^6_t L^2_x} \|P_{N_3} w\|_{L^6_t L^2_x} \\
&\lesssim N_2^{1+} \|P_{N_1} u\|_{X^{0,\frac{1}{3}+}} \|P_{N_2} v\|_{X^{0,\frac{1}{3}+}} \|P_{N_3} w\|_{X^{0,\frac{1}{3}+}}.
\end{aligned} \tag{3.55}$$

This implies the claim by a simple multilinear interpolation with Lemma 3.29; see Exercise 3.16. □

We are now ready to prove:

Theorem 3.31 *For any $s > 0$ with $g \in H^s(\mathbb{T})$, the equation (3.50) is locally wellposed in $X^{s,b}_\delta$ for some $b > \frac{1}{2}$, where $\delta > 0$ depends on $\|g\|_{H^s}$.*

Proof Fix $s > 0$, and $b > \frac{1}{2}$ (it would be chosen sufficiently close to $\frac{1}{2}$). As usual, we will run a fixed point argument for

$$\Gamma u := e^{it\partial_{xx}} g \pm i \int_0^t e^{i(t-t')\partial_{xx}} |u|^4 u dt',$$

on a ball

$$\left\{ u : X^{s,b}_\delta : \|u\|_{X^{s,b}_\delta} \le M \|g\|_{H^s} \right\}.$$

As in the proof of Theorem 3.27, it suffices to prove that

$$\left\| \int_0^t e^{i(t-t')\partial_{xx}} |u|^4 u dt' \right\|_{X^{s,b}_\delta} \lesssim \delta^{0+} \|u\|^5_{X^{s,b}_\delta}.$$

Using Lemma 3.12 and Lemma 3.11, this follows from the inequality

$$\left\| |u|^4 u \right\|_{X^{s,-\frac{1}{2}+}} \lesssim \|u\|^5_{X^{s,\frac{1}{2}-}}. \tag{3.56}$$

By duality, (3.56) follows from

$$\left| \int_{\mathbb{T}\times\mathbb{R}} u\bar{u}u\bar{u}u\bar{h}\, dxdt \right| \lesssim \|u\|^5_{X^{s,\frac{1}{2}-}} \|h\|_{X^{-s,\frac{1}{2}-}}.$$

Consider the contribution of the Littlewood–Paley projections $u_{N_j} := P_{N_j} u$ to the left-hand side

$$\left| \int_{\mathbb{T}\times\mathbb{R}} u_{N_1} \overline{u_{N_2}} u_{N_3} \overline{u_{N_4}} u_{N_5} \overline{h_{N_6}} \, dxdt \right|, \tag{3.57}$$

where dyadic N_j satisfies without loss of generality $N_1 \geq N_2 \geq N_3 \geq N_4 \geq N_5$. Moreover, by L^2 orthogonality, the integral vanishes unless $N_6 \lesssim N_1$.

We consider two cases:

Case 1: $N_6 \approx N_1$.

Using the Cauchy–Schwarz inequality and Corollary 3.30, we have

$$\begin{aligned}
(3.57) &\leq \left\| u_{N_1} u_{N_3} u_{N_5} \right\|_{L^2_{x,t}} \left\| u_{N_2} u_{N_4} h_{N_6} \right\|_{L^2_{x,t}} \\
&\lesssim N_2^{0+} N_3^{0+} \left\| h_{N_6} \right\|_{X^{0,\frac{1}{2}-}} \prod_{j=1}^{5} \left\| u_{N_j} \right\|_{X^{0,\frac{1}{2}-}} \\
&\lesssim N_1^{-s} N_2^{-s+} N_3^{-s+} N_4^{-s} N_5^{-s} N_6^{s} \left\| h_{N_6} \right\|_{X^{-s,\frac{1}{2}-}} \prod_{j=1}^{5} \left\| u_{N_j} \right\|_{X^{s,\frac{1}{2}-}} \\
&\lesssim N_2^{-s+} N_3^{-s+} N_4^{-s} N_5^{-s} \left\| h_{N_6} \right\|_{X^{-s,\frac{1}{2}-}} \prod_{j=1}^{5} \left\| u_{N_j} \right\|_{X^{s,\frac{1}{2}-}}.
\end{aligned}$$

Summing over dyadic N_js using

$$\sum_N \|P_N u\|^2_{X^{s,b}} \lesssim \|u\|^2_{X^{s,b}},$$

we have

$$\begin{aligned}
&\sum_{N_6 \approx N_1 \geq N_2 \geq N_3 \geq N_4 \geq N_5} \left| \int_{\mathbb{T}\times\mathbb{R}} u_{N_1} \overline{u_{N_2}} u_{N_3} \overline{u_{N_4}} u_{N_5} \overline{h_{N_6}} \, dxdt \right| \\
&\lesssim \sum_{N_6 \approx N_1 \geq N_2 \geq N_3 \geq N_4 \geq N_5} N_2^{-s+} N_3^{-s+} N_4^{-s} N_5^{-s} \left\| h_{N_6} \right\|_{X^{-s,\frac{1}{2}-}} \prod_{j=1}^{5} \left\| u_{N_j} \right\|_{X^{s,\frac{1}{2}-}} \\
&\lesssim \|u\|^4_{X^{s,\frac{1}{2}-}} \sqrt{\sum_{N_2,N_3,N_4,N_5} N_2^{-2s+} N_3^{-2s+} N_4^{-2s} N_5^{-2s}} \sum_{N_6 \approx N_1} \left\| h_{N_6} \right\|_{X^{-s,\frac{1}{2}-}} \left\| u_{N_1} \right\|_{X^{s,\frac{1}{2}-}} \\
&\lesssim \|u\|^5_{X^{s,\frac{1}{2}-}} \|h\|_{X^{-s,\frac{1}{2}-}}.
\end{aligned}$$

Case 2: $N_6 \ll N_1$.

In this case, we must have $N_2 \approx N_1$. Using Corollary 3.30 as above, we have

$$(3.57) \lesssim N_3^{0+} \max(N_4^{0+}, N_6^{0+}) \left\|h_{N_6}\right\|_{X^{0,\frac{1}{2}-}} \prod_{j=1}^{5} \left\|u_{N_j}\right\|_{X^{0,\frac{1}{2}-}}$$

$$\lesssim N_1^{-s} N_2^{-s} N_3^{-s+} N_4^{-s+} N_5^{-s} N_6^{s+} \left\|h_{N_6}\right\|_{X^{-s,\frac{1}{2}-}} \prod_{j=1}^{5} \left\|u_{N_j}\right\|_{X^{s,\frac{1}{2}-}}$$

$$\lesssim N_1^{-\frac{s}{2}+} N_2^{-\frac{s}{2}+} N_3^{-s+} N_4^{-s+} N_5^{-s} N_6^{0-} \left\|h_{N_6}\right\|_{X^{-s,\frac{1}{2}-}} \prod_{j=1}^{5} \left\|u_{N_j}\right\|_{X^{s,\frac{1}{2}-}}.$$

Summing in each N_j using Cauchy–Schwarz inequality finishes the proof. □

3.6 Illposedness results

In this section, we present several results on the illposedness of nonlinear dispersive PDEs on $\mathbb{T}$. Similar methods can be used to obtain illposedness results on the real line. We start with cubic NLS on $\mathbb{T}$

$$\begin{cases} iu_t + u_{xx} \pm |u|^2 u = 0, & t \in \mathbb{R},\ x \in \mathbb{T}, \\ u(0,\cdot) = g(\cdot) \in H^s(\mathbb{T}). \end{cases} \tag{3.58}$$

As we have seen in the previous section (3.58) is locally and globally wellposed in $H^s(\mathbb{T})$ for $s \geq 0$. The theory also implies that the data-to-solution map is uniformly continuous on bounded sets. This result is optimal because of the following theorem of Burq–Gérard–Tzvetkov [23]:

Theorem 3.32 *The data-to-solution map of* (3.58) *is not uniformly continuous on bounded balls in H^s for any $s < 0$.*

Proof We only consider the defocusing case (with a minus sign in front of the nonlinearity). The claim for the focusing case follows similarly. Fix $s < 0$ and $a \in \mathbb{R}\backslash\{0\}$. Note that the functions

$$u_{a,n}(t,x) = an^{-s} e^{-it\left(n^2 + a^2 n^{-2s}\right)} e^{inx}$$

solve the equation for each $n \in \mathbb{Z}$ with initial data $u_{a,n}(0,x) = an^{-s}e^{inx}$. Moreover,

$$\|u_{a,n}\|_{H^s} = |a|.$$

Also note that if $a_n \to a$ in $\mathbb{R}$, then

$$\left\|u_{a_n,n}(0,x) - u_{a,n}(0,x)\right\|_{H^s_x} = |a_n - a| \to 0.$$

Fix $t > 0$ and note that

$$\left\|u_{a_n,n}(t,x) - u_{a,n}(t,x)\right\|_{H^s_x} = \left|a_n e^{-ita_n^2 n^{-2s}} - ae^{-ita^2 n^{-2s}}\right|$$
$$\geq |a|\left|e^{-it(a_n^2-a^2)n^{-2s}} - 1\right| - |a_n - a|.$$

Therefore, for $a_n = \sqrt{\pi n^{2s} t^{-1} + a^2}$, we have

$$\left\|u_{a_n,n}(0,x) - u_{a,n}(0,x)\right\|_{H^s_x} \to 0,$$

and

$$\left\|u_{a_n,n}(t,x) - u_{a,n}(t,x)\right\|_{H^s_x} \geq 2|a| - |a_n - a| \to 2|a|,$$

as $n \to \infty$. □

In [39], Christ–Colliander–Tao obtained a stronger illposedness of the equation (3.58) by proving the discontinuity of the flow map in $H^s(\mathbb{T})$, $s < 0$:

Theorem 3.33 *The data-to-solution map of* (3.58) *is not continuous in* $H^s(\mathbb{T})$ *for any* $s < 0$.

We present the proof only for the defocusing case, the focusing case is similar. To prove the theorem, instead of the evolution of a single mode as in Theorem 3.32, we consider two modes: the zero mode and an N mode for some large N. First, one needs to understand the evolution of the initial data $g(x) = a + be^{ix}$ under equation (3.58). One can argue that the solution should remain close to

$$v(t,x) := ae^{-i(|a|^2+2|b|^2)t} + be^{-i(2|a|^2+|b|^2+1)t}e^{ix}, \tag{3.59}$$

by restricting the evolution of (3.58) to the span$\{1, e^{ix}\}$; see Christ–Colliander–Tao [39] and Exercise 3.17. The following justifies this heuristic:

Proposition 3.34 *Fix* $0 < c \ll 1$. *For any* $a, b \in B(0,c) \subset \mathbb{C}$, *the solution* u *of equation* (3.58) *with the initial data* $g(x) = a + be^{ix}$ *satisfies*

$$\sup_{0 \leq t \ll c^{-2}\log(1/c)} \|u(t,\cdot) - v(t,\cdot)\|_{H^1} \lesssim c^3,$$

where v *is as in* (3.59).

Proof First note that

$$iv_t + v_{xx} = a\left(|a|^2 + 2|b|^2\right)e^{-i(|a|^2+2|b|^2)t} + b\left(2|a|^2 + |b|^2\right)e^{-i(2|a|^2+|b|^2+1)t}e^{ix}.$$

Also note that

$$|v|^2 v = a\left(|a|^2 + 2|b|^2\right) e^{-i(|a|^2+2|b|^2)t} + b\left(2|a|^2 + |b|^2\right) e^{-i(2|a|^2+|b|^2+1)t} e^{ix}$$
$$+ a^2\overline{b}e^{i(1-3|b|^2)t}e^{-ix} + b^2\overline{a}e^{-i(2+3|a|^2)t}e^{2ix}$$
$$= iv_t + v_{xx} + a^2\overline{b}e^{i(1-3|b|^2)t}e^{-ix} + b^2\overline{a}e^{-i(2+3|a|^2)t}e^{2ix}.$$

Let $w := u - v$. By plugging $u = v + w$ into (3.58) and using the calculations above, we see that w satisfies

$$\begin{cases} iw_t + w_{xx} = |w + v|^2(w + v) - |v|^2 v + e(t, x), & t \in \mathbb{R}, \ x \in \mathbb{T}, \\ w(0, x) = 0, \end{cases}$$

where

$$e(t, x) = a^2\overline{b}e^{i(1-3|b|^2)t}e^{-ix} + b^2\overline{a}e^{-i(2+3|a|^2)t}e^{2ix}.$$

Since $\|e\|_{H^1} \lesssim c^3$, a direct application of the energy argument gives $\|w\|_{H^1} \lesssim c^3$ only up to $t \approx 1$, which is not good enough. However, observe that one can explicitly solve the equation (see Exersize 3.18)

$$\begin{cases} if_t + f_{xx} = e(t, x), \\ f(0, x) = 0, \end{cases}$$

to obtain

$$f(t, x) = \frac{a^2\overline{b}}{2 - 3|b|^2}\left[e^{it} - e^{i(1-3|b|^2)t}\right]e^{-ix} + \frac{b^2\overline{a}}{2 - 3|a|^2}\left[e^{4it} - e^{-i(2+3|a|^2)t}\right]e^{2ix}.$$

Therefore, recalling that $|a|, |b| \le c \ll 1$, we have

$$\|f\|_{H^1} \lesssim c^3 \quad \text{for all } t.$$

We now set $z = w - f$, which satisfies

$$\begin{cases} iz_t + z_{xx} = |z + f + v|^2(z + f + v) - |v|^2 v := G_{v,f}(z), & t \in \mathbb{R}, \ x \in \mathbb{T}, \\ z(0, x) = 0. \end{cases}$$

Since $u - v = w = z + f$, and $\|f\|_{H^1} \lesssim c^3$ for all times, it suffices to prove that

$$\sup_{0 \le t \ll c^{-2}\log(1/c)} \|z(t, \cdot)\|_{H^1} \lesssim c^3.$$

Note that

$$\partial_t\|z\|^2_{H^1} = -2\Im \int_{\mathbb{T}} \left(z\,\overline{G_{v,f}(z)} + z_x\,\partial_x\overline{G_{v,f}(z)}\right) dx,$$

which implies by the Cauchy–Schwarz inequality

$$\partial_t\|z\|_{H^1} \lesssim \left\|G_{v,f}(z)\right\|_{H^1}.$$

By the algebra property of H^1, and the algebraic inequality

$$\left||\alpha+\beta|^2(\alpha+\beta)-|\beta|^2\beta\right| \lesssim |\alpha|\left(|\alpha|^2+|\beta|^2\right),$$

we obtain

$$\left\|G_{v,f}(z)\right\|_{H^1} \lesssim \|z+f\|_{H^1}\left(\|z+f\|_{H^1}^2+\|v\|_{H^1}^2\right) \lesssim \left(\|z\|_{H^1}+c^3\right)\left(\|z\|_{H^1}^2+c^2\right).$$

In the last inequality, we used the bounds $\|f\|_{H^1} \lesssim c^3$ and $\|v\|_{H^1} \lesssim c$. Let

$$T = \sup\left\{t>0 : \|z\|_{H^1} \le c^3\right\}.$$

On $[0,T]$, we have

$$\partial_t\|z\|_{H^1} \le Cc^5 + Cc^2\|z\|_{H^1}.$$

Therefore, by Gronwall's lemma

$$\|z\|_{H^1} \le Cc^5 e^{Cc^2 t} \le c^3,$$

provided that $t \le T \ll c^{-2}\log(1/c)$. □

Using Proposition 3.34 and a rescaling argument, one can approximate the evolution of $g(x) = a + be^{iNx}$:

Corollary 3.35 *Fix $N \in \mathbb{N}$ and $0 < c \ll N$. For any $a, b \in B(0,c) \subset \mathbb{C}$, the solution u of equation* (3.58) *with the initial data $g(x) = a + be^{iNx}$ satisfies*

$$\sup_{0\le t\ll c^{-2}\log(N/c)} \left|\widehat{u(t)}(0) - \frac{a}{2\pi}e^{-i(|a|^2+2|b|^2)t}\right| \lesssim N^{-2}c^3.$$

Proof First note that u is $\frac{2\pi}{N}$ periodic in x, and that $w(t,x) = \frac{1}{N}u(\frac{t}{N^2}, \frac{x}{N})$ solves the NLS equation with data $\frac{a}{N} + \frac{b}{N}e^{ix}$. Therefore, by Proposition 3.34, for $t \ll N^2c^{-2}\log(N/c)$, we have

$$\left\|w(t,x) - v_{\frac{a}{N},\frac{b}{N}}(t,x)\right\|_{H^1} \lesssim \frac{c^3}{N^3}.$$

Therefore

$$\left|\widehat{w(t)}(0) - \frac{a}{2\pi N}e^{-i(|a|^2+2|b|^2)\frac{t}{N^2}}\right| \lesssim \frac{c^3}{N^3}.$$

Noting that (since u is $\frac{2\pi}{N}$ periodic in x)

$$\mathcal{F}\left(u(t/N^2)\right)(0) = \frac{1}{2\pi}\int_0^{2\pi} u\left(\frac{t}{N^2}, x\right)dx$$
$$= \frac{N}{2\pi}\int_0^{2\pi/N} u\left(\frac{t}{N^2}, x\right)dx = N\widehat{w(t)}(0)$$

yields the claim. □

Proof of Theorem 3.33 We prove the theorem only for $s > -\frac{2}{3}$ and only in the defocusing case. Consider the solution $u = \rho e^{-i|\rho|^2 t}$ of equation (3.58) with initial data $g \equiv \rho$. Also consider the solutions u_N with data

$$g_N(x) = \rho + N^{-s_1} e^{iNx}, \quad N \in \mathbb{N}$$

for some fixed $s_1 \in (s, 0)$. Note that $g_N \to g$ in H^s as $N \to \infty$.

By Corollary 3.35, taking $c = N^{-s_1}$ for large N, and for $t \ll N^{2s_1} \log(N^{1+s_1})$, we have

$$\widehat{u(t)}(0) - \widehat{u_N(t)}(0) = \frac{\rho}{2\pi} e^{-i|\rho|^2 t} - \frac{\rho}{2\pi} e^{-i(|\rho|^2 + 2N^{-2s_1})t} + O(N^{-2-3s_1}).$$

Therefore, for large N

$$\sup_{0<t\ll N^{2s_1}\log(N^{1+s_1})} \left|\widehat{u(t)}(0) - \widehat{u_N(t)}(0)\right|$$

$$= \sup_{0<t\ll N^{2s_1}\log(N^{1+s_1})} \frac{|\rho|}{2\pi}\left|1 - e^{-i2N^{-2s_1}t}\right| + O(N^{-2-3s_1}) \geq \frac{|\rho|}{4\pi}.$$

□

A stronger illposedness result was later given by Molinet in [107]. More precisely, he proved that for any $s < 0$ and for any initial data $g \in H^s(\mathbb{T})$, the data-to-solution map of (3.58) fails to be continuous at g.

Recall from Section 3.5 that the wellposedness theory of the NLS equation on the real line is easier to establish than the theory on the torus. In particular, the result on $\mathbb{R}$ uses only the basic Strichartz estimates, which follows from the $L^1 \to L^\infty$ dispersive estimate; see Chapter 2. In contrast, on $\mathbb{T}$ one needs to prove periodic Strichartz estimates in conjunction with multilinear estimates in the $X^{s,b}$ spaces. By the same token, in the periodic case the absence of dispersion makes the illposedness results easier to establish. We summarize below some of the major illposedness results for the NLS equation on $\mathbb{R}$. In [15] Bourgain proved that below $L^2(\mathbb{R})$ the data-to-solution map is not C^2 (we present the analogous statement and give further discussions for KdV below). The result applies equally well for focusing and defocusing equations. Thus, the equation is illposed below L^2, although the scale invariant Sobolev space is $H^{-\frac{1}{2}}$. In fact, in any dimension and for any power nonlinearity, the local theory cannot go below L^2. On an heuristic level, this is because of the Galilean invariance, which leaves the set of solutions invariant; see Birnir–Kenig–Ponce–Svanstedt–Vega [11]. Indeed, notice that if $u(t,x)$ is a solution (of the focusing equation) with initial data $g(x)$, then

$$u_N(t,x) = e^{-itN^2} e^{iNx} u(t, x - 2tN)$$

is another solution with data $u_N(0,x) = e^{iNx}g(x)$. Thus, if $g \in H^s$ with $s < 0$, and if there is a local existence time $T = T(\|g\|_{H^s})$, then the solution u_N is also wellposed up to T although

$$\lim_{N\to\infty} \|u_N(0,x)\|_{H^s(\mathbb{T})} = 0. \tag{3.60}$$

In this sense, the NLS equation is critical at the L^2 level. Using the Galilean invariance and the existence of solitary wave solutions for the focusing NLS equation, it was proved that the data-to-solution map is not uniformly continuous in $H^s(\mathbb{R})$ for any $s < 0$, [11, 92]. The idea is to consider u_{N_1}, u_{N_2} for a soliton solution u. Because of (3.60), one can make $\|u_{N_1} - u_{N_2}\|_{H^s}$ arbitrarily small by taking N_1 and N_2 large. However, at a later time, the norm of the difference can be bounded below by fixed constant since u_{N_1}, u_{N_2} move with different speeds; see Kenig–Ponce–Vega [92] for details.

In [38], Christ–Colliander–Tao developed a different method and proved the failure of uniform continuity in the range $s \in (-\frac{1}{2}, 0)$ for both focusing and defocusing NLS. Moreover, they demonstrated a more dramatic failure of wellposedness that they called norm inflation in the range $s < -\frac{1}{2}$; see [38] for details.

We now discuss the illposedness theory of the KdV equation on $\mathbb{T}$ (3.23), and related equations. Recall that KdV is wellposed in $H^s(\mathbb{T})$ for $s \geq -\frac{1}{2}$; see Section 3.3.2. It is easy to check that (see Exercise 3.19) the restricted norm method implies the boundedness in $H^s(\mathbb{T})$ of all Picard iterates, in particular the boundedness of the operators:

$$U_1 := W_t g,$$

$$U_2 = \int_0^t W_{t-t'} \partial_x (U_1^2) dt', \tag{3.61}$$

$$U_3 = \int_0^t W_{t-t'} \partial_x (U_1 U_2) dt' = \partial_x \int_0^t W_{t-t'} (U_1 U_2) dt'. \tag{3.62}$$

Note that, if we denote the solution of the KdV equation with the data $\delta g(x)$ by $u(\delta, t, x)$, then using Picard iteration or Duhamel's formula, we have

$$U_j = \frac{\partial^j u(\delta, t, x)}{\partial \delta^j}\Big|_{\delta=0}.$$

Therefore, the following theorem of Bourgain [18] proves that the restricted norm method we applied in Section 3.3.2 (or other contraction arguments) cannot be extended to the range $s < -\frac{1}{2}$. Moreover, the data-to-solution map cannot be C^3.

Theorem 3.36 *The operator $g \to U_3$ is not bounded in $H^s(\mathbb{T})$ for any $s < -\frac{1}{2}$.*

Proof Fix N large and consider the data

$$g(x) = N^{-s}\cos(Nx).$$

Note that the H^s norm of g is 1. We claim that the H^s norm of U_3 is $\gtrsim tN^{-1-2s}$, which is unbounded as $N \to \infty$ for any fixed $s < -\frac{1}{2}$ and $t > 0$. To prove the claim, we calculate

$$U_1 = N^{-s}\cos(Nx + N^3 t),$$

$$\begin{aligned} U_2 = -2N^{1-2s} &\int_0^t W_{t-t'}[\sin(Nx + N^3 t')\cos(Nx + N^3 t')]dt' \\ &= -N^{1-2s}\int_0^t W_{t-t'}[\sin(2Nx + 2N^3 t')]dt' \\ &= -N^{1-2s}\int_0^t \sin(2Nx + 2N^3 t' + 8N^3(t - t'))dt' \\ &= -\frac{1}{6}N^{-2-2s}[\cos(2Nx + 2N^3 t) - \cos(2Nx + 8N^3 t)]. \end{aligned}$$

Note that up to a constant

$$\begin{aligned} U_1 U_2 = N^{-2-3s}\Big[&\cos\left(Nx + N^3 t\right) - \cos\left(Nx + 7N^3 t\right) \\ &+ \cos\left(3Nx + 3N^3 t\right) - \cos\left(3Nx + 9N^3 t\right)\Big]. \end{aligned}$$

Therefore

$$\begin{aligned} W_{t-t'}(U_1 U_2) = N^{-2-3s}\Big[&\cos\left(Nx + N^3 t\right) - \cos\left(Nx + 7N^3 t - 6N^3 t'\right) \\ &+ \cos\left(3Nx + 27N^3 t - 24N^3 t'\right) - \cos\left(3Nx + 27N^3 t - 18N^3 t'\right)\Big]. \end{aligned}$$

Note that the only t'-independent term is the first one; the others contribute negative powers of N after the integration in t'. Therefore, for fixed $t > 0$ and for sufficiently large N, it suffices to consider the contribution of the first term. This leads to

$$\|U_3\|_{H^s} \approx \left\|\int_0^t W_{t-t'}(U_1 U_2)dt'\right\|_{H^{s+1}} \gtrsim tN^{-2-3s}N^{s+1} = tN^{-1-2s},$$

which is unbounded for $s < -\frac{1}{2}$. □

Recall that the wellposedness theory established by Kappeler–Topalov in [81] establishes a continuous flow in $H^{-1}(\mathbb{T})$. This result is optimal since in [110] Molinet proved that the flow map is not continuous in $H^s(\mathbb{T})$ for any $s < -1$.

Before we present other illposedness results on $\mathbb{T}$, we briefly discuss the KdV equation on $\mathbb{R}$. Bourgain's theorem above can be extended to the case of the real line establishing that the iterative methods cannot be used in the range $s < -\frac{3}{4}$, and in particular the data-to-solution map cannot be C^2; see Bourgain [18] and Tzvetkov [152]. In [38], Christ–Colliander–Tao proved that the data-to-solution map cannot be uniformly continuous in $H^s(\mathbb{R})$ for $-1 < s < -\frac{3}{4}$. Finally, in [109] Molinet proved that the data-to-solution map fails to be continuous in $H^s(\mathbb{R})$ for any $s < -1$.

We note that, unlike the NLS equation and the KdV equation on $\mathbb{R}$, the lack of uniform continuity is not a good indicator of illposedness for the KdV equation on $\mathbb{T}$. The wellposedness theory (see Section 3.3.2) implies the uniform continuity of the evolution only on the bounded subsets of

$$H_c^s(\mathbb{T}) = \left\{ g \in H^s(\mathbb{T}) : \frac{1}{2\pi} \int_{\mathbb{T}} g = c \right\},$$

for any fixed $s \geq -\frac{1}{2}$ and $c \in \mathbb{R}$. However, the uniform continuity fails on bounded subsets of $H^s(\mathbb{T})$ for any s, although we have well-defined continuous solutions that exist for all times; see Molinet [106]. The reason for this is the Galilean invariance, which is the fact that if u solves the equation on $\mathbb{T}$, then

$$v(t, x) = \omega + u(t, x - \omega t)$$

is also a solution for any $\omega \in \mathbb{R}$. Indeed, at time zero u and v differ only by a (small) constant; however, since they rotate with different frequencies, they separate at later times. In this section, we present the proof in the range $s \in (0, 1)$:

Theorem 3.37 *For $s \in (0, 1)$, the KdV evolution* (3.23) *is not uniformly continuous on any ball in $H^s(\mathbb{T})$.*

Proof We prove the statement only for the unit ball in $H^s(\mathbb{T})$. Fix $N \in \mathbb{N}$ and consider the following solution of the Airy equation

$$\phi(t, x) = N^{-s} \sin(Nx + N^3 t)$$

with initial data $g(x) = N^{-s} \sin(Nx)$. Note that $\|\phi\|_{H^r} = N^{r-s}$, for all r. Let u be the solution of the equation (3.23) with initial data $g(x)$. By the energy conservation law, $\|u\|_{H^1} \lesssim N^{1-s}$.

Claim: For $t \lesssim N^{s-1}$, we have $\|u - \phi\|_{L^2} \lesssim t N^{1-2s}$.

Indeed, letting $v := u - \phi$ we see that v satisfies

$$\begin{cases} v_t + v_{xxx} + v v_x + \partial_x(v\phi) + \frac{1}{2} N^{1-2s} \sin(2Nx + 2N^3 t) = 0, \\ v(0, x) = 0. \end{cases}$$

Therefore, by integration by parts

$$\partial_t\|v\|_{L^2}^2 = -\int_0^{2\pi} v^2\phi_x dx + N^{1-2s}\int_0^{2\pi} v\sin(2Nx+2N^3t)dx$$
$$\lesssim \|\phi_x\|_{L^\infty}\|v\|_{L^2}^2 + N^{1-2s}\|v\|_{L^2}$$
$$\lesssim N^{1-s}\|v\|_{L^2}^2 + N^{1-2s}\|v\|_{L^2}.$$

Thus

$$\partial_t\|v\|_{L^2} \lesssim N^{1-s}\|v\|_{L^2} + N^{1-2s}, \quad \|v(0)\|_{L^2} = 0,$$

which implies, by Gronwall's lemma, that (for $t \lesssim N^{s-1}$)

$$\|v\|_{L^2} \lesssim N^{1-2s}\int_0^t e^{N^{1-s}\tau}d\tau \lesssim tN^{1-2s}.$$

Interpolating the L^2-bound with the bound

$$\|v\|_{H^1} \le \|u\|_{H^1} + \|\phi\|_{H^1} \lesssim N^{1-s},$$

we obtain (for $t \lesssim N^{s-1}$)

$$\|v\|_{H^s} \lesssim \|v\|_{L^2}^{1-s}\|v\|_{H^1}^{s} \lesssim t^{1-s}N^{(1-s)^2}. \tag{3.63}$$

Pick $a \in (1-s, 1)$ and let

$$t_N = N^{-a}, \quad \omega_1 = -\omega_2 = \frac{\pi}{2Nt_N}.$$

Consider the following solutions of equation (3.23)

$$u_j(t,x) = \omega_j + u(t, x-\omega_j t), \quad j = 1,2.$$

We have

$$\|u_1(0,x) - u_2(0,x)\|_{H^s} = |\omega_1 - \omega_2| \lesssim N^{-1+a} = o(1),$$

as $N \to \infty$. Also note that

$$\|u_1(t_N,x) - u_2(t_N,x)\|_{H^s_x}$$
$$\ge \|u(t_N, x-\omega_1 t_N) - u(t_N, x-\omega_2 t_N)\|_{H^s_x} - |\omega_1 - \omega_2|$$
$$\ge \left\|\phi\left(t_N, x-\frac{\pi}{2N}\right) - \phi\left(t_N, x+\frac{\pi}{2N}\right)\right\|_{H^s_x} - CN^{-a(1-s)}N^{(1-s)^2} - N^{-1+a}$$
$$= \left\|\phi\left(t_N, x-\frac{\pi}{2N}\right) - \phi\left(t_N, x+\frac{\pi}{2N}\right)\right\|_{H^s_x} + o(1).$$

In the second inequality, we used the bound (3.63).

Now note that

$$\phi\left(t_N, x - \frac{\pi}{2N}\right) - \phi\left(t_N, x + \frac{\pi}{2N}\right)$$
$$= N^{-s}\left(\sin(Nx + N^3 t_N - \frac{\pi}{2}) - \sin(Nx + N^3 t_N + \frac{\pi}{2})\right)$$
$$= -2N^{-s}\cos(Nx + N^3 t_N),$$

whose H^s norm is 2. Therefore

$$\|u_1(t_N, x) - u_2(t_N, x)\|_{H^s_x} \geq 1$$

for large N. □

We should note that before the result in [106] the authors in [38] have proved failure of uniform continuity of the data-to-solution map in $H^s(\mathbb{T})$ for any $s \in (-2, -\frac{1}{2})$.

We finish this chapter with a discussion of the family of equations:

$$u_t - D^\alpha u_x + uu_x = 0, \quad t \in \mathbb{R}, \quad x \in \mathbb{T} \text{ or } \mathbb{R}, \tag{3.64}$$

for any $0 < \alpha < 2$. The case $\alpha = 2$ corresponds to the KdV equation, which we already discussed. When $\alpha = 0$, the equation is the well-known Burgers' equation. The case $\alpha = 1$ corresponds to the Benjamin–Ono (BO) equation. First observe that the problem is scale invariant at the $H^{\frac{1}{2}-\alpha}$ level. The failure of analytic wellposedness in $H^s(\mathbb{R})$ is known for any $s \in \mathbb{R}$ and $0 < \alpha < 2$, see Molinet–Saut–Tzvetkov [112]. More precisely, it was proven that the data the solution map is not C^2 in $H^s(\mathbb{R})$. In [104], Linares–Pilod–Saut, using the technique in [11], proved that the data-to-solution map fails to be uniformly continuous at the critical regularity $H^{\frac{1}{2}-\alpha}(\mathbb{R})$ for any $\frac{1}{3} \leq \alpha \leq \frac{1}{2}$. The restriction on α comes from the fact that solitary wave solutions of equation (3.64) are known to exist only in this range of α.

There are more illposedness results in the special case of the BO equation, which is scale invariant at the $H^{-\frac{1}{2}}$ level. For $s < -\frac{1}{2}$, it is known that the data-to-solution map is not uniformly continuous in $H^s(\mathbb{R})$, see Biagioni–Linares [10]. By using the conservation of the mean, it was proved in Koch–Tzvetkov [97] that the same result holds for any $s > 0$. The method used in [97] inspired the illposedness result (Theorem 3.37) in [106]. For the periodic BO, the method presented in the proof of Theorem 3.36 yields that the data-to-solution map fails to be C^3 for any $s < 0$. Moreover, the illposedness result in [106] remains valid for any $\alpha \in (0, 2)$, in particular for the BO equation. Later, the failure of continuity in $H^s(\mathbb{T})$ for any $s < 0$ was given in Molinet [108]. If one assumes mean-zero initial data then illposedness results, especially for the periodic problem, are harder to establish. Nevertheless, in Molinet [111] it was proved that the data-to-solution map fails to be $C^{1+\rho}$ in $H^s_0(\mathbb{T})$ for any $\rho > 0$ and $s < 0$.

We now discuss the wellposedness theory of (3.64). For any $\alpha \geq 0$, a refinement of the energy method presented in Section 3.1 gives local wellposedness for any $s > \frac{3}{2}$ for both the real and the periodic case. For the BO equation on $\mathbb{R}$, recall that wellposedness cannot be established by iteration methods [112]. Major progress on the problem was made in [142], where Tao proved that the BO equation is globally wellposed in $H^1(\mathbb{R})$. This was achieved by performing a global gauge transformation which almost entirely eliminates the derivative in the nonlinearity. This technique was pushed further by Ionescu–Kenig in [77] where global wellposedness was established in $L^2(\mathbb{R})$. The method can be adapted to the periodic case. Indeed in [111], Molinet obtained the global wellposedness in $L^2(\mathbb{T})$ which is sharp because of the result in [108].

Finally, for $0 < \alpha < 1$, Linares–Pilod–Saut [104] used a refinement of the oscillatory integral method along with the energy method and established local wellposedness in $H^s(\mathbb{R})$ for any $s > \frac{3}{2} - \frac{3\alpha}{8}$. For wellposedness results in the range $1 < \alpha < 2$, see the work of Guo in [73], and the references therein.

Exercises

3.1 (Duhamel's principle) Let L be linear symmetric differential operator with constant coefficients. Let $u \in C^1_t C^\infty_x([0,T] \times \mathbb{T})$ and that $F \in C^0_t C^\infty_x([0,T] \times \mathbb{T})$. Then u solves

$$\begin{cases} iu_t - Lu = F, & x \in \mathbb{T}, \quad t \in [0,T], \\ u(0,x) = g(x) \in C^\infty, \end{cases} \tag{3.65}$$

if and only if

$$u(t,x) = e^{-itL} g(x) - i \int_0^t e^{-i(t-s)L} F(s,x) ds, \; t \in [0,T].$$

3.2 Suppose that we have a dispersive PDE with L^2-norm conservation. Fix $s > 0$. Prove that if the local existence time δ in H^s depends only on the L^2 norm of the initial data, then the following global bound holds

$$\|u(t)\|_{H^s} \leq C e^{C|t|} \|u(0)\|_{H^s}.$$

3.3 Prove (3.7) using the commutator estimates in Lemma 1.9.

3.4 Fix $s \in \mathbb{R}$. Prove that

$$\|V\|_{Y^s_\delta} \lesssim \|V\|_{C^2_t H^{s+\frac{3}{2}}_x}$$

uniformly in $\delta > 0$.

3.5 Use the algebra property of $H^s(\mathbb{T})$, $s > \frac{1}{2}$ and Duhamel's principle to prove that (3.48) is locally wellposed in $C_t^0 H_x^s$.

3.6 Prove, using the Gagliardo–Nirenberg inequality, that the smooth solutions of (3.48) satisfy

$$\|u\|_{H^1(\mathbb{T})} \leq C = C(\|g\|_{H^1}),$$

both in the focusing and defocusing cases.

3.7 For $s \geq 1$, prove that

$$\left\||u|^2 u\right\|_{H^s(\mathbb{T})} \lesssim \|u\|_{H^1(\mathbb{T})}^2 \|u\|_{H^s(\mathbb{T})}.$$

3.8 Prove using Exercise 3.6, Exercise 3.7, and Gronwall's inequality that the smooth solutions of the cubic NLS equation (3.48) satisfy for $s \geq 1$

$$\|u\|_{H^s(\mathbb{T})} \leq \|g\|_{H^s(\mathbb{T})} e^{Ct}.$$

Conclude that (3.48) is globally wellposed on $H^s(\mathbb{T})$, $s \geq 1$.

3.9 Prove Theorem 3.25 by modifying the proof of Theorem 3.18.

3.10 In this exercise, we consider the L^2 solutions of the mass subcritical NLS on $\mathbb{R}^n$ (see Tsutsumi [151])

$$\begin{cases} iu_t + \Delta u + |u|^{p-1} u = 0, & x \in \mathbb{R}^n,\ t \in \mathbb{R}, \\ u(0, \cdot) = g(\cdot) \in L^2(\mathbb{R}^n), & 1 < p < 1 + \frac{4}{n}, \end{cases} \tag{3.66}$$

and prove that they satisfy the equation in the distributional sense (see Stein–Shackarchi [134]).

(a) Show that

$$e^{it\Delta} g \in C_t^0 L_x^2(\mathbb{R} \times \mathbb{R}^n)$$

provided that $\|g\|_{L^2(\mathbb{R}^n)} < \infty$.

(b) Show that if $g \in L^2(\mathbb{R})$, then $e^{it\Delta} g$ satisfies the linear Schrödinger equation

$$iu_t + \Delta u = 0$$

in the sense of space-time distributions.

(c) Assume $F \in C_0^\infty(\mathbb{R} \times \mathbb{R}^n)$. By Exercise 3.1

$$D(F)(t, x) := -i \int_0^t e^{i(t-s)\Delta} F(s, x) ds$$

satisfies (3.65) with zero initial data. Show that $D(F)$ is a C^∞ function. Then show that

$$\|D(F)(t)\|_{L_x^2} \leq |t|^{\frac{1}{2}} \|F\|_{L_t^2 L_x^2},$$

and that $D(F) \in C_t^0 L_x^2(\mathbb{R} \times \mathbb{R}^n)$.

(d) Use Strichartz estimates (Exercise 2.1) to prove that

$$D(F) \in C_t^0 L_x^2(\mathbb{R} \times \mathbb{R}^n)$$

when $F \in L_t^{q'} L_x^{r'}$ for all dual exponents (q', r'), of admissible (q, r).

(e) Now assume that u solves

$$u(t,x) = e^{it\Delta} g - i \int_0^t e^{i(t-s)\Delta} |u|^{p-1} u(s) ds$$

in $C_t^0 L_x^2 \cap L_t^q L_x^r$ with $p < 1 + \frac{4}{n}$. Then show that u solves (3.66) in the sense of distributions.

3.11 Consider the KdV equation

$$u_t + u_{xxx} + u u_x = 0, \quad x \in \mathbb{R},\ t \in \mathbb{R}. \tag{3.67}$$

Construct soliton solutions of the form $u(t,x) = f(x - ct)$, where f satisfies

$$f(x), f'(x), f''(x) \to 0 \text{ as } |x| \to \infty,$$

by following the steps:

(a) Obtain a second-order equation for f by plugging $u(t,x) = f(x - ct)$ into the equation and then integrating.

(b) Multiply the resulting equation with f' and integrate the result.

(c) Set $g^2 = 3c - f$, assuming $0 < f < 3c$ to get a first-order equation for g.

3.12 Prove the following inequalities assuming that $\alpha + \beta > 1$ and $\alpha \geq \beta \geq 0$

$$\int_{\mathbb{R}} \frac{1}{\langle x - k_1 \rangle^\alpha \langle x - k_2 \rangle^\beta} dx \lesssim \langle k_1 - k_2 \rangle^{-\beta} \phi_\alpha(k_1 - k_2), \text{ and}$$

$$\sum_{n \in \mathbb{Z}} \frac{1}{\langle n - k_1 \rangle^\alpha \langle n - k_2 \rangle^\beta} \lesssim \langle k_1 - k_2 \rangle^{-\beta} \phi_\alpha(k_1 - k_2),$$

where

$$\phi_\alpha(k) := \sum_{|n| \leq |k|} \frac{1}{\langle n \rangle^\alpha} \sim \begin{cases} 1, & \alpha > 1, \\ \log(1 + \langle k \rangle), & \alpha = 1, \\ \langle k \rangle^{1-\beta}, & \alpha < 1. \end{cases}$$

3.13 For $\beta > 1$, prove that

$$\int_{\mathbb{R}} \frac{1}{\langle x \rangle^\beta \sqrt{|x - y|}} dx \lesssim \frac{1}{\langle y \rangle^{1/2}}.$$

3.14 Let Q be the unique positive solution (see Berestycki–Lions [5] and Kwong [101]) of the equation

$$Q_{xx} + Q^5 = Q.$$

Prove that the function

$$u(x,t) = |t|^{-\frac{1}{2}} e^{i\frac{x^2-4}{4t}} Q\left(\frac{x}{t}\right)$$

solves the focusing quintic NLS equation (3.47) on $\mathbb{R}$ for $t \neq 0$.

3.15 Prove inequality (3.51) using Theorem 2.10.

3.16 Prove the inequality (3.54) by interpolating the inequalities (3.53) and (3.55).
Hint: This follows from linear interpolation iteratively in u, v, and w by fixing the remaining functions.

3.17 Prove that the function v given in (3.59) solves the cubic NLS equation (3.58) restricted to the span of $\{1, e^{ix}\}$ by writing the equation on the Fourier side, and considering only the first two modes.

3.18 Show that the solution of the equation

$$\begin{cases} if_t + f_{xx} = a^2\overline{b}e^{i(1-3|b|^2)t}e^{-ix} + b^2\overline{a}e^{-i(2+3|a|^2)t}e^{2ix}, \\ f(0,x) = 0 \end{cases}$$

is given by

$$f(t,x) = \frac{a^2\overline{b}}{2-3|b|^2}\left[e^{it} - e^{i(1-3|b|^2)t}\right]e^{-ix} + \frac{b^2\overline{a}}{2-3|a|^2}\left[e^{4it} - e^{-i(2+3|a|^2)t}\right]e^{2ix}.$$

3.19 Prove that the operators defined in (3.61) and (3.62) are bounded in $H^s(\mathbb{T})$ for $s > -\frac{1}{2}$ using the results of Section 3.3.2.

4

Global dynamics of nonlinear dispersive PDEs

In this chapter, we study the regularity properties of the solutions of NLS and KdV type equations on the torus. In the first section, we present a smoothing property of the data-to-solution map, which states that the nonlinear part of the solution is smoother than the initial data; see Christ [37], Erdoğan–Tzirakis [56, 57, 59, 60], Kappeler-Schaad–Topalov [79, 80], Oh [119], Demirbaş–Erdoğan–Tzirakis [51], and Compaan [47]. For the cubic NLS equation, the smoothing property we present follows from a direct trilinear estimate in $X^{s,b}$ spaces. This method fails for the KdV equation. To obtain the smoothing property, one needs to apply a normal form transform as in Section 3.4, and then prove a trilinear estimate for the modified nonlinearity.

In the second section, we present Bourgain's high–low decomposition method [19], which was the first technique employed to obtain global solutions for infinite energy initial data. In [19], the high–low decomposition method relied on one-sided smoothing estimates for the nonlinearity. We present the method in the case of the KdV equation with a potential using the smoothing result from the first section. The smoothing property simplifies the presentation; however, the potential term causes additional difficulties since there are no conserved quantities anymore.

Finally, in the third section, we present another globalizing technique, the "I–method", which is a refinement of the high–low decomposition method based on the construction of almost conserved quantities. This idea first appeared in Keel–Tao [86]. Later it was developed further by Colliander–Keel–Staffilani–Takaoka–Tao; see e.g. [41, 42, 43, 44]. We present this technique for the quintic NLS equation on the torus, which is L^2-critical and known to be locally wellposed in $H^s(\mathbb{T})$ for $s > 0$; see Section 3.5.

4.1 Smoothing for nonlinear dispersive PDEs on the torus

4.1.1 Cubic NLS on the torus

We start with the cubic NLS equation on the torus

$$\begin{cases} iu_t + u_{xx} + |u|^2 u = 0, & x \in \mathbb{T}, \quad t \in \mathbb{R}, \\ u(0,\cdot) = g(\cdot) \in H^s(\mathbb{T}), \end{cases} \tag{4.1}$$

Recall that the equation (4.1) is locally and globally wellposed in H^s for $s \geq 0$; see Bourgain [15] and Section 3.5. Also recall that the corresponding $X^{s,b}$ space is defined via the norm

$$\|u\|_{X^{s,b}} = \left\|e^{-it\partial_{xx}}u\right\|_{H^s_x H^b_t} = \left\|\langle k\rangle^s \left\langle \tau + k^2\right\rangle^b \widehat{u}(\tau,k)\right\|_{L^2_\tau \ell^2_k}.$$

The restricted norm $X^{s,b}_\delta$ is defined analogously, and the lemmas 3.9, 3.10, 3.11, and 3.12 that we proved in Section 3.3 remain valid. Our aim is to prove the following theorem from [59]:

Theorem 4.1 *Fix $s > 0$. Let u be the solution to* (4.1) *with data $g \in H^s(\mathbb{T})$. Then for any $a < \min(2s, \frac{1}{2})$, we have*

$$u(t,x) - e^{it(\partial_{xx}+P)}g \in C^0_t H^{s+a}_x(\mathbb{R}\times\mathbb{T}),$$

where $P := \frac{1}{\pi}\|g\|^2_{L^2}$.

In [37], Christ obtained a (local-in-time) smoothing estimate in the $\mathcal{F}\ell^p_s \to \mathcal{F}\ell^q_s$ setting for the cubic NLS equation. Here $\mathcal{F}\ell^p_s$ is the space of functions whose Fourier series is in

$$\ell^p_s = \left\{(a_n)_{n\in\mathbb{Z}} : \sum_n |a_n|^p \langle n\rangle^{ps} < \infty\right\}.$$

In [80], Kappeler–Schaad–Topalov obtained a smoothing statement as above for the defocusing cubic NLS equation for $s \geq 2$, $s \in \mathbb{N}$, and $a \leq 1$. They also obtained a stronger smoothing statement with a modified linear group for $s \geq 1$ so that the difference in the linear and nonlinear evolutions remains in a bounded subset of H^{s+1}. This proves that if the initial data is in H^s, $s \geq 1$, then the solution remains in a bounded subset of H^s. This statement follows from the conserved quantities for integer values of $s \geq 0$. It remains open for $0 < s < 1$.

There are many smoothing results as in Theorem 4.1 for dispersive PDEs on $\mathbb{R}^n$, partially relying on the smoothing properties of the linear group. The first such result was obtained by Bourgain in [19] for the cubic NLS equation on $\mathbb{R}^2$ as a by-product of the high–low decomposition method; see Keraani–Vargas [93] for related results in other dimensions; also Fonseca–Linares–Ponce [62, 63] for related results on the generalized KdV equation on $\mathbb{R}$. However, it appears that the KdV equation on the real line does not have this smoothing property; see Colliander–Staffilani–Takaoka [46] for a smoothing result in Sobolev norms that suppresses frequencies around the origin; also Exercise 4.8 for a smoothing statement for the KdV equation on the real line with special initial data.

We start the proof of Theorem 4.1 with the change of variable $u(t,x) \to u(t,x)e^{iPt}$, which leads to the equation

$$\begin{cases} iu_t + u_{xx} + |u|^2 u - Pu = 0, \quad x \in \mathbb{T},\ t \in \mathbb{R} \\ u(0,x) = g(x). \end{cases} \tag{4.2}$$

It suffices to prove that the solution of this modified equation satisfies

$$u - e^{it\partial_{xx}} g \in C_t^0 H_x^{s+a}(\mathbb{R} \times \mathbb{T}). \tag{4.3}$$

Using Plancherel and the conservation of the L^2-norm, we rewrite

$$\begin{aligned} \widehat{|u|^2 u}(k) - P\widehat{u}(k) &= -P\widehat{u}(k) + \sum_{k_1,k_2} \widehat{u}(k_1)\overline{\widehat{u}(k_2)}\widehat{u}(k - k_1 + k_2) \\ = -P\widehat{u}(k) &+ \frac{1}{\pi}\|u\|_2^2 \widehat{u}(k) - |\widehat{u}(k)|^2 \widehat{u}(k) + \sum_{k_1 \neq k, k_2 \neq k_1} \widehat{u}(k_1)\overline{\widehat{u}(k_2)}\widehat{u}(k - k_1 + k_2) \\ &= -|\widehat{u}(k)|^2 \widehat{u}(k) + \sum_{k_1 \neq k, k_2 \neq k_1} \widehat{u}(k_1)\overline{\widehat{u}(k_2)}\widehat{u}(k - k_1 + k_2) \\ &=: \widehat{\rho(u)}(k) + \widehat{R(u)}(k). \end{aligned} \tag{4.4}$$

We decompose the solution of (4.2) as

$$u(t,x) = e^{it\partial_{xx}} g + \mathcal{N}(t,x),$$

where $\mathcal{N}$ is the nonresonant part of the nonlinear Duhamel term of the solution

$$\mathcal{N}(t,x) = \int_0^t e^{i(t-s)\partial_{xx}} \Big(\rho(u)(s,x) + R(u)(s,x)\Big) ds.$$

Observe that the case $a = 0$ of the following lemma gives an alternative proof of the local wellposedness in $H^s(\mathbb{T})$ for $s > 0$. Note that the local existence at the L^2 level and control of the $X^{s,b}$ norm of the solution in the local existence interval were given in Theorem 3.27. Also recall from Exercise 3.8 that, for any $s \geq 0$, we have the following a priori bound

$$\|u\|_{H^s} \leq Ce^{C|t|}\|g\|_{H^s} =: T(t). \tag{4.5}$$

The smoothing property of the NLS equation relies on the following multilinear smoothing estimate for the nonlinearity in $X^{s,b}$ spaces.

Lemma 4.2 *For fixed $s > 0$, $\delta > 0$, and $a < \min(2s, \frac{1}{2})$, we have*

$$\|R(u) + \rho(u)\|_{X_\delta^{s+a,b-1}} \lesssim \|u\|^3_{X_\delta^{s,b}}$$

provided that $0 < b - \frac{1}{2}$ is sufficiently small.

Proof As usual, we ignore the δ dependence. We start with the ρ term. Note that

$$\|\rho\|_{X^{s+a,b-1}}$$
$$= \left\| \langle k\rangle^{s+a} \langle \tau + k^2\rangle^{b-1} \int_{\mathbb{R}^2} \widehat{u}(\tau_1, k)\overline{\widehat{u}(-\tau_2, -k)}\widehat{u}(\tau - \tau_1 - \tau_2, k) d\tau_1 d\tau_2 \right\|_{\ell_k^2 L_\tau^2}.$$

Letting

$$f(\tau, k) = |\widehat{u}(\tau, k)| \langle k\rangle^s \left\langle \tau + k^2 \right\rangle^b,$$

it suffices to prove that

$$\left\| \int_{\mathbb{R}^2} \frac{\langle k\rangle^{-2s+a} \left\langle \tau + k^2\right\rangle^{b-1} f(\tau_1, k) f(-\tau_2, -k) f(\tau - \tau_1 - \tau_2, k) d\tau_1 d\tau_2}{\langle \tau_1 + k^2\rangle^b \langle \tau_2 - k^2\rangle^b \langle \tau - \tau_1 - \tau_2 + k^2\rangle^b} \right\|_{\ell_k^2 L_\tau^2} \lesssim \|f\|^3_{\ell_k^2 L_\tau^2}.$$

Using $a < 2s$ and the Cauchy–Schwarz inequality in the τ_1, τ_2 integrals, we estimate the square of the left-hand side by

$$\sup_{\tau,k} \left(\int_{\mathbb{R}^2} \frac{\left\langle \tau + k^2 \right\rangle^{2b-2}}{\langle \tau_1 + k^2\rangle^{2b} \langle \tau_2 - k^2\rangle^{2b} \langle \tau - \tau_1 - \tau_2 + k^2\rangle^{2b}} d\tau_1 d\tau_2 \right) \times$$
$$\sum_k \int_{\mathbb{R}^3} f^2(\tau_1, k) f^2(-\tau_2, -k) f^2(\tau - \tau_1 - \tau_2, k) d\tau_1 d\tau_2 d\tau.$$

Using Exercise 3.12 in τ_1 and then in τ_2 integrals inside the supremum, and integrating in τ, τ_1, τ_2 in the second line, we bound this by

$$\sum_k \|f(\tau, k)\|^6_{L_\tau^2} \leq \left[\sum_k \|f(\tau, k)\|^2_{L_\tau^2} \right]^3 = \|f\|^6_{\ell_k^2 L_\tau^2}.$$

Now, we consider the R term (after a change of variable $(\tau_2, k_2) \to (-\tau_2, -k_2)$)

$$\|R(u)\|^2_{X^{s+a,b-1}}$$
$$= \left\| \int_{\tau_1,\tau_2} \sum_{k_1 \neq k, k_2 \neq k_1} \frac{\langle k \rangle^{s+a} \widehat{u}(\tau_1, k_1) \overline{\widehat{u}(\tau_2, k_2)} \widehat{u}(\tau - \tau_1 + \tau_2, k - k_1 + k_2)}{\langle \tau + k^2 \rangle^{1-b}} \right\|^2_{\ell^2_k L^2_\tau}.$$

Defining f as above, it suffices to prove that

$$\left\| \int_{\mathbb{R}^2} \sum_{k_1 \neq k, k_2 \neq k_1} M\, f(\tau_1, k_1) f(\tau_2, k_2) f(\tau - \tau_1 + \tau_2, k - k_1 + k_2) d\tau_1 d\tau_2 \right\|^2_{\ell^2_k L^2_\tau} \lesssim \|f\|_2^6,$$

where

$$M = M(k_1, k_2, k, \tau_1, \tau_2, \tau) =$$
$$\frac{\langle k \rangle^{s+a} \langle k_1 \rangle^{-s} \langle k_2 \rangle^{-s} \langle k - k_1 + k_2 \rangle^{-s}}{\langle \tau + k^2 \rangle^{1-b} \left\langle \tau_1 + k_1^2 \right\rangle^b \left\langle \tau_2 + k_2^2 \right\rangle^b \langle \tau - \tau_1 + \tau_2 + (k - k_1 + k_2)^2 \rangle^b}. \quad (4.6)$$

By the Cauchy–Schwarz inequality in τ_1, τ_2, k_1, k_2 variables, we estimate the norm above by

$$\sup_{k,\tau} \left(\int \sum_{k_1 \neq k, k_2 \neq k_1} M^2(k_1, k_2, k, \tau_1, \tau_2, \tau) d\tau_1 d\tau_2 \right) \times$$
$$\left\| \int \sum_{k_1, k_2} f^2(\tau_1, k_1) f^2(\tau_2, k_2) f^2(\tau - \tau_1 + \tau_2, k - k_1 + k_2) d\tau_1 d\tau_2 \right\|_{\ell^1_k L^1_\tau}.$$

Note that the norm above is equal to $\left\| f^2 * f^2 * f^2 \right\|_{\ell^1_k L^1_\tau}$, which can be estimated by $\|f\|^6_{\ell^2_k L^2_\tau}$ by Young's inequality. Therefore, it suffices to prove that the supremum above is finite. Using Exercise 3.12 in the τ_1 and τ_2 integrals, we obtain

$$\sup_{k,\tau} \int \sum_{k_1 \neq k, k_2 \neq k_1} M^2 d\tau_1 d\tau_2$$
$$\lesssim \sup_{k,\tau} \sum_{k_1 \neq k, k_2 \neq k_1} \frac{\langle k \rangle^{2s+2a} \langle k_1 \rangle^{-2s} \langle k_2 \rangle^{-2s} \langle k - k_1 + k_2 \rangle^{-2s}}{\langle \tau + k^2 \rangle^{2-2b} \left\langle \tau + k_1^2 - k_2^2 + (k - k_1 + k_2)^2 \right\rangle^{4b-1}}$$
$$\lesssim \sup_{k} \sum_{k_1 \neq k, k_2 \neq k_1} \frac{\langle k \rangle^{2s+2a} \langle k_1 \rangle^{-2s} \langle k_2 \rangle^{-2s} \langle k - k_1 + k_2 \rangle^{-2s}}{\left\langle k^2 - k_1^2 + k_2^2 - (k - k_1 + k_2)^2 \right\rangle^{2-2b}}.$$

The last line follows by the simple fact

$$\langle \tau - n\rangle\langle \tau - m\rangle \gtrsim \langle n - m\rangle. \tag{4.7}$$

Since we have only the nonresonant terms, $k_1 \neq k, k_2$, we have

$$\left\langle k^2 - k_1^2 + k_2^2 - (k - k_1 + k_2)^2 \right\rangle = \langle (k - k_1)(k_1 - k_2)\rangle \approx \langle k - k_1\rangle\langle k_1 - k_2\rangle.$$

Therefore, it suffices to estimate

$$\sum_{k_1,k_2} \frac{\langle k\rangle^{2s+2a}\langle k_1\rangle^{-2s}\langle k_2\rangle^{-2s}\langle k - k_1 + k_2\rangle^{-2s}}{\langle k - k_1\rangle^{2-2b}\langle k_1 - k_2\rangle^{2-2b}}.$$

To estimate this sum, we consider the cases $|k - k_1 + k_2| \gtrsim |k|$, $|k_1| \gtrsim |k|$, and $|k_2| \gtrsim |k|$. In the estimate, we always take $0 < b - \frac{1}{2}$ sufficiently small for any given $s > 0$ and $0 \leq a < \min(2s, \frac{1}{2})$.

In the first case, using Exercise 3.12 in the k_2 sum we bound the contribution of these terms by

$$\sum_{k_1,k_2} \frac{\langle k\rangle^{2a}\langle k_1\rangle^{-2s}\langle k_2\rangle^{-2s}}{\langle k - k_1\rangle^{2-2b}\langle k_1 - k_2\rangle^{2-2b}} \lesssim \sum_{k_1} \frac{\langle k\rangle^{2a}\phi_{\max(2s,2-2b)}(k_1)}{\langle k - k_1\rangle^{2-2b}\langle k_1\rangle^{2s+\min(2-2b,2s)}}.$$

In the case $s \geq \frac{1}{2}$, we bound the sum by

$$\sum_{k_1} \frac{\langle k\rangle^{2a}\log(1 + \langle k_1\rangle)}{\langle k - k_1\rangle^{2-2b}\langle k_1\rangle^{2s+2-2b}} \lesssim \langle k\rangle^{2a-2+2b+} \lesssim 1$$

provided $a < \frac{1}{2}$.

In the case $0 < s < \frac{1}{2}$, using Exercise 3.12 we bound the sum by

$$\sum_{k_1} \frac{\langle k\rangle^{2a}}{\langle k - k_1\rangle^{2-2b}\langle k_1\rangle^{4s+1-2b}} \lesssim \begin{cases} \langle k\rangle^{2a+4b-4s-2} & 0 < s \leq \frac{1}{4}, \\ \langle k\rangle^{2a+2b-2} & \frac{1}{4} < s < \frac{1}{2}. \end{cases}$$

This is bounded in k provided that $0 \leq a < \min(2s, \frac{1}{2})$.

The second case is identical to the first case after renaming the variables: $n_1 = k - k_1 + k_2$, $n_2 = k_2$.

In the third case, after renaming the variables $n_1 = k_1$, $n_2 = k - k_1 + k_2$, and using Exercise 3.12 we bound the sum by

$$\sum_{n_1,n_2} \frac{\langle k\rangle^{2a}\langle n_1\rangle^{-2s}\langle n_2\rangle^{-2s}}{\langle k - n_1\rangle^{2-2b}\langle k - n_2\rangle^{2-2b}} \lesssim \langle k\rangle^{2a}\phi^2_{\max(2s,2-2b)}(k)\langle k\rangle^{-2\min(2-2b,2s)} \lesssim 1$$

provided that $s > 0$ and $0 \leq a < \min(2s, 1)$. □

Using Lemma 3.9, Lemma 3.12, and Lemma 4.2, we obtain (in the local existence interval $[-\delta, \delta]$)

$$\|u - e^{it\partial_{xx}} g\|_{C_t^0 H_x^{s+a}([-\delta,\delta]\times\mathbb{T})} \lesssim \|\mathcal{N}\|_{X_\delta^{s+a,b}}$$
$$\lesssim \|\rho + R\|_{X_\delta^{s,b-1}} \lesssim \|u\|^3_{X_\delta^{s,b}} \lesssim \|g\|^3_{H^s}. \quad (4.8)$$

This implies (4.3) and hence Theorem 4.1 in the local existence interval $[-\delta, \delta]$. To prove Theorem 4.1 for all times, we need to iterate this local result. Fix t large. For $r \leq t$, we have the bound (see (4.5))

$$\|u(r)\|_{H^s} \lesssim T(r) \leq T(t).$$

Applying (4.8), we have

$$\left\|u(j\delta) - e^{i\delta\partial_{xx}} u((j-1)\delta)\right\|_{H^{s+a}} \lesssim \|u((j-1)\delta)\|^3_{H^s} \lesssim T(t)^3,$$

for any $j \leq t/\delta$. Using this, we obtain (with $J = t/\delta$)

$$\left\|u(J\delta) - e^{iJ\delta\partial_{xx}} g\right\|_{H^{s+a}} = \left\|\sum_{j=1}^{J} e^{i(J-j)\delta\partial_{xx}} u(j\delta) - e^{i(J-j+1)\delta\partial_{xx}} u((j-1)\delta)\right\|_{H^{s+a}}$$
$$\leq \sum_{j=1}^{J} \left\|e^{i(J-j)\delta\partial_{xx}} u(j\delta) - e^{i(J-j+1)\delta\partial_{xx}} u((j-1)\delta)\right\|_{H^{s+a}}$$
$$= \sum_{j=1}^{J} \left\|u(j\delta) - e^{i\delta\partial_{xx}} u((j-1)\delta)\right\|_{H^{s+a}}$$
$$\lesssim JT(t)^3 \approx \frac{tT(t)^3}{\delta}.$$

This finishes the proof of Theorem 4.1.

4.1.2 The KdV equation on the torus

In this section, we establish the smoothing property for forced and damped KdV equation. We start by noting that for any $s \in \mathbb{R}$ and for any $a > 0$, the inequality

$$\|(uv)_x\|_{X^{s+a,-1/2}} \lesssim \|u\|_{X^{s,1/2}} \|v\|_{X^{s,1/2}} \quad (4.9)$$

fails,[1] where $X^{s,b}$ is the Bourgain space associated with the KdV equation; see Exercise 4.2. This suggests that the smoothing property we established above for the NLS equation does not hold for the KdV equation if one views

[1] However, this method works for the KdV type equations with higher order dispersion; see Exercise 4.4

the nonlinear evolution as a perturbation of the linear flow and apply standard Picard iteration techniques to absorb the nonlinear derivative term. However, the first Picard iteration for the KdV equation has the correction term

$$e^{-t\partial_x^3}\int_0^t e^{t'\partial_x^3}\left[e^{-t'\partial_x^3}g\partial_x\left(e^{-t'\partial_x^3}g\right)\right]dt'.$$

This suggests a full derivative smoothing for the nonlinearity, since on the Fourier side (ignoring zero modes)

$$\sum_{k_1+k_2=k}\int_0^t e^{-3ik_1k_2kt'}k_2\widehat{g}(k_1)\widehat{g}(k_2)dt' = \sum_{k_1+k_2=k}\frac{\widehat{g}(k_1)\widehat{g}(k_2)}{-3ikk_1}\left(e^{-3ik_1k_2kt}-1\right).$$

Therefore, if $g \in L^2$, then the correction term is in H^1 (see Exercise 4.3).

To obtain the smoothing property, we apply a normal form transform as in Section 3.4, and then prove a trilinear smoothing estimate for the modified nonlinearity instead of (4.9).

We establish the smoothing result in the more general setting of the forced and damped KdV with a potential

$$\begin{cases} u_t + (\partial_{xxx}+\gamma)u + \left(Vu + \frac{u^2}{2}\right)_x = f, \quad x \in \mathbb{T},\ t \in \mathbb{R}, \\ u(0,x) = g(x) \in H^s(\mathbb{T}). \end{cases} \tag{4.10}$$

Here $\gamma \geq 0$, $f \in C_t^2 H_x^{s+\frac{3}{2}}(\mathbb{R}\times\mathbb{T})$, and $V \in C_t^2 H_x^{s+3}(\mathbb{R}\times\mathbb{T})$. Moreover, g is mean-zero, and V and f are mean-zero for each t. We work with this generality for various applications of the smoothing bound that will be discussed in the later sections. Recall that we already discussed the local wellposedness theory in Section 3.3.3. The mean-zero assumption on the initial data can be removed easily in the case of $\gamma = 0$, as remarked in Section 3.3.3.

Theorem 4.3 *Fix $s > -1/2$ and $a < \min(2s+1, 1)$. Consider the real valued solution of the KdV equation (4.10) on $\mathbb{R}\times\mathbb{T}$ with mean-zero initial data $u(0,x) = g(x) \in H^s(\mathbb{T})$. Assume that we have an a priori growth bound*

$$\|u(t)\|_{H^s} \leq C(\|g\|_{H^s})T(t)$$

for some nondecreasing function $T(t)$. Then

$$u(t) - W_t^\gamma g \in C_t^0 H_x^{s+a},$$

where $W_t^\gamma = e^{-t\partial_{xxx}-t\gamma}$. Moreover, we have

$$\left\|u(t) - W_t^\gamma g\right\|_{H^{s+a}} \leq C(s,a,\|g\|_{H^s},\gamma,V,f)T(t)^3\min\left(\frac{1}{1-e^{-\delta\gamma}}, \langle t\rangle T(t)^{6+}\right).$$

This theorem was proved in [56, 57]. More recently, Kappeler–Schaad–Topalov [79] improved this result using inverse scattering methods in the case $V = f = \gamma = 0$ and for nonnegative integers s. They proved that

$$\|u(t) - W_t g\|_{H^{s+1}} \lesssim \langle t \rangle C(\|g\|_{H^s}),$$

where $W_t = W_t^0$. They also obtained a stronger version with a modified linear group as we discussed above.

We now discuss the method of the proof from [56, 57]. Following the argument in Babin–Ilyin–Titi [1] (see Section 3.4), we write the equation on the Fourier side and use differentiation by parts to take advantage of the large denominators that appear due to the dispersion. In this particular form, the derivative in the nonlinearity is eliminated. The penalty one pays after such a reduction is an increase in the order of the nonlinearity (from quadratic to cubic in the case of the KdV equation), and the appearance of the second order resonant terms. Due to the absence of the zero Fourier mode, the bilinear nonlinearity has no resonant terms. To estimate the new trilinear term, we now decompose the nonlinearity into resonant and nonresonant terms. It should be noted that in the resonant terms the waves interact with no oscillation and hence they are always "the enemy." However, it turns out that the nonsmooth resonant terms of the KdV equation cancel out. Furthermore, the gain of the derivative is more than enough to compensate for the remaining nonlinear terms. For the nonresonant terms, we apply the restricted norm method of Bourgain (see Section 3.3) to the reduced nonlinearity to prove the theorem. We note that a similar combination of integration by parts and the restricted norm method was used by Takaoka–Tsutsumi in [139].

Using the notation

$$u(t,x) = \sum_k u_k(t)e^{ikx}, \quad V(t,x) = \sum_k V_k(t)e^{ikx}, \text{ and } f(t,x) = \sum_k f_k(t)e^{ikx},$$

we write (4.10) on the Fourier side

$$\begin{cases} \partial_t u_k = \left(ik^3 - \gamma\right) u_k - \frac{ik}{2} \sum_{k_1+k_2=k} (2V_{k_1} + u_{k_1})\, u_{k_2} + f_k, \\ u_k(0) = \widehat{g}(k). \end{cases}$$

Because of the mean-zero assumption on u, V, and f, there are no zero harmonics in this equation. Using the transformations

$$\begin{aligned} u_k(t) &= v_k(t)e^{(ik^3-\gamma)t}, \\ V_k(t) &= \frac{1}{2}\Lambda_k(t)e^{(ik^3-\gamma)t}, \\ f_k(t) &= h_k(t)e^{(ik^3-\gamma)t}, \end{aligned}$$

and the identity

$$(k_1 + k_2)^3 - k_1^3 - k_2^3 = 3(k_1 + k_2)k_1k_2,$$

the equation can be written in the form

$$\partial_t v_k = -\frac{ik}{2}e^{-\gamma t} \sum_{k_1+k_2=k} e^{-3ikk_1k_2t}(\Lambda_{k_1} + v_{k_1})v_{k_2} + h_k(t). \tag{4.11}$$

We need the following proposition which follows from differentiation by parts:

Proposition 4.4 *The system* (4.11) *can be written in the following form*

$$\partial_t \left(v_k + e^{-\gamma t}B_2(v + \Lambda, v)_k\right) = -\gamma e^{-\gamma t}B_2(v + \Lambda - \frac{\Lambda_t}{\gamma}, v)_k + h_k(t)$$
$$+ e^{-\gamma t}B_2(h, 2v + \Lambda)_k + e^{-2\gamma t}\rho_k + e^{-2\gamma t}D_3(2v + \Lambda, v + \Lambda, v)_k, \tag{4.12}$$

where we define

$$B_2(f, g)_0 = \rho_0 = D_3(f, g, h)_0 = 0,$$

and for $k \neq 0$, we define

$$B_2(f, g)_k = -\frac{1}{6} \sum_{k_1+k_2=k} \frac{e^{-3ikk_1k_2t} f_{k_1}g_{k_2}}{k_1k_2}$$

$$\rho_k = \frac{i}{12}\Lambda_k \sum_{|j|\neq|k|} \frac{\Lambda_j\overline{v_j}}{j} - \frac{i}{12}\frac{(\overline{\Lambda_k} + 2\overline{v_k})(\Lambda_k + v_k)v_k}{k}$$

$$D_3(f, g, h)_k = \frac{i}{12} \sum_{\substack{k_1+k_2+k_3=k \\ (k_1+k_2)(k_1+k_3)(k_2+k_3)\neq 0}} \frac{e^{-3it(k_1+k_2)(k_2+k_3)(k_3+k_1)}}{k_1} f_{k_1}g_{k_2}h_{k_3}.$$

Proof By differentiation by parts (see Section 3.4), we rewrite (4.11) as follows

$$\partial_t v_k = -e^{-\gamma t}\partial_t B_2(v + \Lambda, v)_k$$
$$+ e^{-\gamma t}B_2\left((v + \Lambda)_t, v\right)_k + e^{-\gamma t}B_2(v + \Lambda, v_t)_k + h_k(t).$$

Commuting ∂_t and $e^{-\gamma t}$ and using the symmetry and bilinearity of the operator B, we obtain

$$\partial_t \left(v_k + e^{-\gamma t}B_2(v + \Lambda, v)_k\right)$$
$$= -\gamma e^{-\gamma t}B_2\left(v + \Lambda - \frac{\Lambda_t}{\gamma}, v\right)_k + e^{-\gamma t}B_2\left(v_t, 2v + \Lambda\right)_k + h_k(t).$$

Using the formula (4.11) for v_t, we obtain (see Section 4.4 for details)

$$\partial_t \Big(v_k + e^{-\gamma t} B_2(v + \Lambda, v)_k\Big)$$
$$= -\gamma e^{-\gamma t} B_2\left(v + \Lambda - \frac{\Lambda_t}{\gamma}, v\right)_k + h_k(t) + e^{-\gamma t} B_2(h, 2v + \Lambda)$$
$$+ \frac{ie^{-2\gamma t}}{12} \sum_{\substack{k_1+k_2+k_3=k \\ k_2+k_3\neq 0}} \frac{e^{-3it(k_1+k_2)(k_2+k_3)(k_3+k_1)}}{k_1} (2v + \Lambda)_{k_1} (v + \Lambda)_{k_2} v_{k_3}.$$

To finish the proof, it suffices to consider the contribution of the resonant terms (3.36) in the last summand

$$\sum_{\ell=1}^{3} \sum_{R_\ell} \frac{(2v + \Lambda)_{k_1} (v + \Lambda)_{k_2} v_{k_3}}{k_1}.$$

Recall from Section 3.4 that

$$R_1 = \{k_1 + k_2 = 0\} \cap \{k_3 + k_1 = 0\} \Leftrightarrow \{k_1 = -k,\ k_2 = k,\ k_3 = k\},$$

$$R_2 = \{k_1 + k_2 = 0\} \cap \{k_3 + k_1 \neq 0\} \Leftrightarrow \{k_1 = j,\ k_2 = -j,\ k_3 = k,\ |j| \neq |k|\},$$

$$R_3 = \{k_3 + k_1 = 0\} \cap \{k_1 + k_2 \neq 0\}\} \Leftrightarrow \{k_1 = j,\ k_2 = k,\ k_3 = -j,\ |j| \neq |k|\}.$$

The proposition follows if we show that the formula above is equal to $-12i\rho_k$. Note that

$$\sum_{\ell=1}^{3} \sum_{R_\ell} \frac{(\Lambda_{k_1} + 2v_{k_1})(\Lambda_{k_2} + v_{k_2}) v_{k_3}}{k_1} = -\frac{(\Lambda_{-k} + 2v_{-k})(\Lambda_k + v_k) v_k}{k} \tag{4.13}$$
$$+ v_k \sum_{|j|\neq|k|} \frac{(\Lambda_j + 2v_j)(\Lambda_{-j} + v_{-j})}{j} + (\Lambda_k + v_k) \sum_{|j|\neq|k|} \frac{(\Lambda_j + 2v_j) v_{-j}}{j}$$

Using $v_j = \overline{v_{-j}}$ and $\Lambda_j = \overline{\Lambda_{-j}}$, we can rewrite the second line above as

$$v_k \sum_{|j|\neq|k|} \frac{|\Lambda_j + v_j|^2 + |v_j|^2 + v_j \overline{\Lambda_j}}{j} + (\Lambda_k + v_k) \sum_{|j|\neq|k|} \frac{(\Lambda_j \overline{v_j} + 2|v_j|^2)}{j}$$
$$= v_k \sum_{|j|\neq|k|} \frac{v_j \overline{\Lambda_j}}{j} + (\Lambda_k + v_k) \sum_{|j|\neq|k|} \frac{\Lambda_j \overline{v_j}}{j}$$
$$= 2v_k \sum_{|j|\neq|k|} \frac{\Re(v_j \overline{\Lambda_j})}{j} + \Lambda_k \sum_{|j|\neq|k|} \frac{\Lambda_j \overline{v_j}}{j}.$$

The first equality follows from the symmetry relation $j \leftrightarrow -j$. By the same token, the first summand in the last line above vanishes since

$$\Re(v_j \overline{\Lambda_j}) = \Re(\overline{v_j} \Lambda_j) = \Re(v_{-j} \overline{\Lambda_{-j}}).$$

Using this in (4.13), we obtain

$$\sum_{\ell=1}^{3} \sum_{R_\ell} \frac{(\Lambda_{k_1} + 2v_{k_1})(\Lambda_{k_2} + v_{k_2})v_{k_3}}{k_1} = -\frac{(\Lambda_{-k} + 2v_{-k})(\Lambda_k + v_k)v_k}{k} + \Lambda_k \sum_{|j|\neq|k|} \frac{\Lambda_j \overline{v_j}}{j} = \frac{12}{i}\rho_k,$$

which yields the assertion of the proposition. □

Integrating (4.12) from 0 to t, we obtain

$$\begin{aligned} v_k(t) - v_k(0) &= -e^{-\gamma t} B_2(\Lambda + v, v)_k(t) + B_2(\Lambda + v, v)_k(0) \\ &- \int_0^t e^{-\gamma r} B_2(\gamma v + \gamma\Lambda - \partial_r \Lambda, v)_k(r) dr + \int_0^t e^{-\gamma r} B_2(h, 2v + \Lambda)_k(r) dr \\ &+ \int_0^t \left(h_k(r) + e^{-2\gamma r}\rho_k(r)\right) dr + \int_0^t e^{-2\gamma r} D_3(\Lambda + 2v, \Lambda + v, v)_k(r) dr. \end{aligned}$$

Transforming back to the u, V, f functions, we have

$$\begin{aligned} u_k(t) - e^{ik^3 t - \gamma t} g_k &= -\mathcal{B}(2V + u, u)_k(t) + e^{ik^3 t - \gamma t}\mathcal{B}(2V + g, g)_k(0) \\ &+ \int_0^t e^{ik^3(t-r)} e^{-\gamma(t-r)} \left(\mathcal{B}\left(-\gamma u + 2e^{rL}\partial_r(e^{-rL}V), u\right)_k(r) + 2\mathcal{B}(f, u + V)_k(r)\right) dr \\ &+ \int_0^t e^{ik^3(t-r)} e^{-\gamma(t-r)} \left(f_k(r) + \tilde{\rho}_k(r)\right) dr \\ &+ \int_0^t e^{ik^3(t-r)} e^{-\gamma(t-r)} \mathcal{D}(V + u, 2V + u, u)_k(r) dr, \end{aligned} \tag{4.14}$$

where

$$\begin{aligned} \mathcal{B}(f, g)_k &= -\frac{1}{6} \sum_{k_1 + k_2 = k} \frac{f_{k_1} g_{k_2}}{k_1 k_2}, \\ \tilde{\rho}_k &= \frac{i}{3} V_k \sum_{|j|\neq|k|} \frac{V_j \overline{u_j}}{j} - \frac{i}{6} \frac{(\overline{V_k} + \overline{u_k})(2V_k + u_k)u_k}{k} \\ \mathcal{D}(f, g, h)_k &= \frac{i}{6} \sum_{\substack{k_1 + k_2 + k_3 = k \\ (k_2 + k_3)(k_1 + k_2)(k_1 + k_3) \neq 0}} \frac{f_{k_1} g_{k_2} h_{k_3}}{k_1}. \end{aligned}$$

We can rewrite this on the space side as

$$
\begin{aligned}
u(t) - W_t^\gamma g = &-\mathcal{B}(2V + u, u)(t) + W_t^\gamma \mathcal{B}(2V(0) + g, g) \\
&+ \int_0^t W_{t-r}^\gamma \left(\mathcal{B}(-\gamma u + 2W_r \partial_r (W_{-r} V), u)(r) + 2\mathcal{B}(f, u + V)(r)\right) dr \\
&+ \int_0^t W_{t-r}^\gamma (f(r) + \tilde{\rho}(r))\, dr + \int_0^t W_{t-r}^\gamma \mathcal{D}(V + u, 2V + u, u)(r) dr.
\end{aligned} \tag{4.15}
$$

Lemma 4.5 *For $s > -\frac{1}{2}$ and $a \le 1$, we have*

$$\|\mathcal{B}(u, v)\|_{H^{s+a}} \lesssim \|u\|_{H^s} \|v\|_{H^s}.$$

For $s > -\frac{1}{2}$ and $0 \le a \le 2s + 1$, we have

$$\|\tilde{\rho}\|_{H^{s+a}} \lesssim \|u\|_{H^s} \left(\|V\|_{H^{s+a}}^2 + \|u\|_{H^s}^2\right).$$

Proof In the estimate for $\mathcal{B}(u, v)$, we can assume by symmetry that $|k_1| \gtrsim |k|$. Thus, for $a \le 1$ and $s > -\frac{1}{2}$, we have

$$
\begin{aligned}
\|\mathcal{B}(u, v)\|_{H^{s+a}} &\lesssim \Big\| \sum_{k_1 + k_2 = k,\ |k_1| \gtrsim |k|} \frac{|k|^{s+a} |u_{k_1} v_{k_2}|}{|k_1 k_2|} \Big\|_{\ell_k^2} \\
&\lesssim \Big\| \sum_{k_1 + k_2 = k,\ |k_1| \ge |k_2|} \frac{|k_1|^s |u_{k_1}| |v_{k_2}|}{|k_2|} \Big\|_{\ell_k^2} \lesssim \Big\| \frac{v_k}{k} \Big\|_{\ell^1} \big\| |k|^s u_k \big\|_{\ell^2} \\
&\lesssim \big\| |k|^s v_k \big\|_{\ell^2} \big\| |k|^{-1-s} \big\|_{\ell^2} \|u\|_{H^s} \lesssim \|u\|_{H^s} \|v\|_{H^s}.
\end{aligned}
$$

In the last line, we used Young's and Cauchy–Schwarz inequalities, and the fact that $s > -\frac{1}{2}$.

Now note that for $a \le 2s + 1$

$$
\begin{aligned}
\Big\| \frac{u_k v_k w_k}{k} |k|^{s+a} \Big\|_{\ell^2} &= \big\| u_k v_k w_k |k|^{3s} |k|^{a-2s-1} \big\|_{\ell^2} \\
&\le \big\| u_k v_k w_k |k|^{3s} \big\|_{\ell^2} \lesssim \|u\|_{H^s} \|v\|_{H^s} \|w\|_{H^s}.
\end{aligned}
$$

In the last inequality, we used $\ell^2 \subset \ell^\infty$. Also note that for any $a \ge 0$

$$
\begin{aligned}
\Bigg\| V_k \sum_{|j| \ne |k|} \frac{|V_j| |v_j|}{|j|} |k|^{s+a} \Bigg\|_{\ell^2} &\le \|V\|_{H^{s+a}} \sum_j \frac{|V_j|}{|j|^{1+s}} |u_j| |j|^s \\
&\le \|V\|_{H^{s+a}} \|V\|_{H^{-s-1}} \|u\|_{H^s} \le \|V\|_{H^{s+a}}^2 \|u\|_{H^s}.
\end{aligned}
$$

The last two estimates imply the bound for $\tilde{\rho}$. □

Using the estimates in Lemma 4.5 and

$$\|W_r \partial_r (W_{-r} V)\|_{H^s} \lesssim \|\partial_r V - LV\|_{H^s} \lesssim \|\partial_r V\|_{H^s} + \|V\|_{H^{s+3}},$$

we bound the right-hand side of (4.15) (for $s > -1/2$ and $a < \min(2s+1, 1)$) to obtain

$$\left\|u(t) - W_t^\gamma g\right\|_{H^{s+a}} \lesssim \|2V(t) + u(t)\|_{H^s}\|u(t)\|_{H^s} + \|2V(0) + g\|_{H^s}\|g\|_{H^s}$$
$$+ \int_0^t e^{-\gamma(t-r)}\|u(r)\|_{H^s}\left(\|u(r)\|_{H^s} + \|\partial_r V\|_{H^s} + \|V(r)\|_{H^{s+3}}\right) dr$$
$$+ \int_0^t e^{-\gamma(t-r)}\|f(r)\|_{H^s}\left(\|u(r)\|_{H^s} + \|V(r)\|_{H^s}\right) dr$$
$$+ \int_0^t e^{-\gamma(t-r)}\left(\|f(r)\|_{H^{s+a}} + \|u(r)\|_{H^s}\left(\|V(r)\|^2_{H^{s+a}} + \|u(r)\|^2_{H^s}\right)\right) dr$$
$$+ \left\|\int_0^t W_{t-r}^\gamma \mathcal{D}(V+u, 2V+u, u)(r) dr\right\|_{H^{s+a}}.$$

And hence

$$\left\|u(t) - W_t^\gamma g\right\|_{H^{s+a}} \lesssim \|u(t)\|_{H^s}\|V(t)\|_{H^s} + \|u(t)\|^2_{H^s} + \|g\|_{H^s}\|V(0)\|_{H^s} + \|g\|^2_{H^s}$$
$$+ \int_0^t e^{-\gamma(t-r)}\left(1 + \|u(r)\|_{H^s}\right)\left(1 + \|u(r)\|^2_{H^s} + \|V\|^2_{C_r^1 H_x^{s+3}} + \|f(r)\|^2_{H^{s+a}}\right) dr$$
$$+ \left\|\int_0^t W_{t-r}^\gamma \mathcal{D}(V+u, 2V+u, u)(r) dr\right\|_{H^{s+a}}. \quad (4.16)$$

Since our nonlinearity after differentiation by parts is not uu_x anymore, we can avoid the Y^{s+a} and Z^{s+a} spaces that we needed to establish the local theory; see Section 3.3.2. Instead we use the embedding $X_\delta^{s+a,b} \subset C_t^0 H_x^{s+a}([-\delta, \delta] \times \mathbb{T})$ for $b > \frac{1}{2}$ and the following lemma.

Lemma 4.6 *Let $-\frac{1}{2} < b' \leq 0$ and $b = b' + 1$. Then for any $\delta < 1$ and $\gamma \geq 0$*

$$\left\|\int_0^t W_{t-r}^\gamma F(r)\, dr\right\|_{X_\delta^{s,b}} \lesssim \|F\|_{X_\delta^{s,b'}}, \quad (4.17)$$

where the implicit constant depends on γ and b.

Proof We will prove this for $\gamma > 0$ since the case $\gamma = 0$ is in Section 3.3.1, see Lemma 3.12.

Let η be a smooth function supported on $[-2, 2]$ and $\eta(t) = 1$ for $|t| \leq 1$. It suffices to prove the statement with $X^{s,b}$ norms

$$\left\|\eta(t)\int_0^t W_{t-r}^\gamma F(r)\, dr\right\|_{X^{s,b}} = \left\|\eta(t)\int_0^t e^{-\gamma(t-r)} W_{-r} F(r)\, dr\right\|_{H_x^s H_t^b}.$$

Therefore, as in the proof of Lemma 3.12, it suffices to prove that

$$\left\|\eta(t)\int_0^t e^{-\gamma(t-r)} f(r)\, dr\right\|_{H^b} \lesssim \|f\|_{H^{b'}}. \quad (4.18)$$

Writing

$$\int_0^t e^{-\gamma(t-r)} f(r)dr = \int [\chi_{[0,t]} e^{-\gamma(t-r)}]^\vee(z)\widehat{f}(z)dz = \int \frac{e^{izt} - e^{-\gamma t}}{\gamma + iz}\widehat{f}(z)dz,$$

we see that the Fourier transform of the function inside the norm in the left-hand side of (4.18) is

$$\int \frac{\widehat{\eta}(\tau - z) - \widehat{\eta e^{-\gamma \cdot}}(\tau)}{\gamma + iz}\widehat{f}(z)dz.$$

For the contribution of this to the left-hand side of (4.18), we use the inequalities

$$\langle \tau \rangle^b \lesssim \langle \tau - z \rangle^b \langle z \rangle^b, \qquad \frac{1}{|\gamma + iz|} \lesssim \frac{1}{\langle z \rangle},$$

and Young's inequality to get

$$\begin{aligned}
&\left\| \langle \tau \rangle^b \int \frac{\widehat{\eta}(\tau - z) - \widehat{\eta e^{-\gamma \cdot}}(\tau)}{\gamma + iz}\widehat{f}(z)dz \right\|_{L^2} \\
&\quad \lesssim \left\| \langle \tau \rangle^b \int \frac{|\widehat{\eta}(\tau - z)| + |\widehat{\eta e^{-\gamma \cdot}}(\tau)|}{\langle z \rangle} \left|\widehat{f}(z)\right| dz \right\|_{L^2} \\
&\quad \lesssim \left\| \int \left(\frac{\langle \tau - z \rangle^b |\widehat{\eta}(\tau - z)|}{\langle z \rangle^{1-b}} + \frac{\langle \tau \rangle^b |\widehat{\eta e^{-\gamma \cdot}}(\tau)|}{\langle z \rangle} \right) \left|\widehat{f}(z)\right| dz \right\|_{L^2} \\
&\quad \lesssim \left\| \langle \tau \rangle^b \widehat{\eta} \right\|_{L^1} \left\| \langle z \rangle^{b-1} \widehat{f}(z) \right\|_{L^2} + \left\| \langle \tau \rangle^b \widehat{\eta e^{-\gamma \cdot}} \right\|_{L^2} \left\| \langle z \rangle^{-1} \widehat{f}(z) \right\|_{L^1} \\
&\quad \lesssim \left\| \langle z \rangle^{b'} \widehat{f}(z) \right\|_{L^2} + \left\| \langle z \rangle^{b'} \widehat{f}(z) \right\|_{L^2} \left\| \langle z \rangle^{-1-b'} \right\|_{L^2} \lesssim \|f\|_{H^{b'}}.
\end{aligned}$$

The forth inequality holds since $\eta(t)e^{-\gamma t}$ is a Schwartz function. The last inequality follows from the fact that $-1 - b' < -1/2$. □

For $|t| < \delta$, where δ as in Theorem 3.21, and $b > \frac{1}{2}$, we have

$$\begin{aligned}
&\left\| \int_0^t W_{t-r}^\gamma \mathcal{D}(V + u, 2V + u, u)(r)dr \right\|_{H^{s+a}} \\
&\quad \leq \left\| \eta(t) \int_0^t W_{t-r}^\gamma \mathcal{D}(V + u, 2V + u, u)(r)\, dr \right\|_{L^\infty_{t\in[-\delta,\delta]} H_x^{s+a}} \\
&\quad \lesssim \left\| \eta(t) \int_0^t W_{t-r}^\gamma \mathcal{D}(V + u, 2V + u, u)(r)\, dr \right\|_{X_\delta^{s+a,b}} \\
&\quad \lesssim \|\mathcal{D}(V + u, 2V + u, u)\|_{X_\delta^{s+a,b-1}}. \qquad (4.19)
\end{aligned}$$

Proposition 4.7 *For $s > -\frac{1}{2}$, $a < \min(1, 2s+1)$, and for $0 < b - \frac{1}{2}$ sufficiently small, we have*

$$\|\mathcal{D}(u, v, w)\|_{X_\delta^{s+a,b-1}} \leq C\|u\|_{X_\delta^{s,1/2}}\|v\|_{X_\delta^{s,1/2}}\|w\|_{X_\delta^{s,1/2}}.$$

We will prove this proposition later on. Using (4.19) and the proposition above in (4.16), we see that for $|t| < \delta$ we have

$$\begin{aligned}\|u(t) - W_t^\gamma g\|_{H^{s+a}} &\lesssim \|u(t)\|_{H^s}\|V(t)\|_{H^s} + \|u(t)\|_{H^s}^2 + \|g\|_{H^s}\|V(0)\|_{H^s} + \|g\|_{H^s}^2 \\ &+ \int_0^t e^{-\gamma(t-r)}\,(1 + \|u(r)\|_{H^s})\left(1 + \|u(r)\|_{H^s}^2 + \|V(r)\|_{H^{s+1}}^2 + \|f(r)\|_{H^{s+a}}^2\right) dr \\ &+ \|u\|_{X_\delta^{s,\frac{1}{2}}}^3 + \|u\|_{X_\delta^{s,\frac{1}{2}}}\|V\|_{X_\delta^{s,\frac{1}{2}}}^2.\end{aligned}$$

In the rest of the proof, the implicit constants also depend on $\|g\|_{H^s}$, V and f. Fix t large. For $r \leq t$, assume the bound

$$\|u(r)\|_{H^s} \lesssim T(r) \leq T(t).$$

We also assume that $T(t) \geq 1$ without loss of generality. Thus, by the local theory, with $\delta \approx T(t)^{-6-}$, we have

$$\left\|u(j\delta) - W_\delta^\gamma u\,((j-1)\delta)\right\|_{H^{s+a}} \lesssim T(t)^3,$$

for any $j \leq t/\delta \lesssim \langle t\rangle T(t)^6$. Here we used the local theory bound

$$\|u\|_{X^{s,1/2}_{[(j-1)\delta,\, j\delta]}} \lesssim \|u((j-1)\delta)\|_{H^s} \lesssim T(t).$$

Using this, we obtain (with $J = t/\delta \lesssim \langle t\rangle T(t)^{6+}$)

$$\begin{aligned}\left\|u(t) - W_t^\gamma g\right\|_{H^{s+a}} &= \left\|u(J\delta) - W_{J\delta}^\gamma g\right\|_{H^{s+a}} \\ &\leq \sum_{j=1}^{J}\left\|W_{(J-j)\delta}^\gamma u(j\delta) - W_{(J-j+1)\delta}^\gamma u((j-1)\delta)\right\|_{H^{s+a}} \\ &= \sum_{j=1}^{J} e^{-(J-j)\delta\gamma}\left\|u(j\delta) - W_\delta^\gamma u\,((j-1)\delta)\right\|_{H^{s+a}} \lesssim T(t)^3\sum_{j=1}^{J} e^{-(J-j)\delta\gamma} \\ &\lesssim T(t)^3\min\left(\frac{1}{1-e^{-\delta\gamma}}, \langle t\rangle T(t)^{6+}\right).\end{aligned}$$

This completes the proof of the growth bound stated in Theorem 4.3. The continuity part of the theorem follows from the embedding of $X^{s+a,b}$ into $C_t^0 H_x^{s+a}$, the continuity of V and f, and the a priori bound on $\mathcal{B}$.

4.1.3 Proof of Proposition 4.7

Recall that

$$\mathcal{D}(u,v,w)(r,x) = \sum_{k\neq 0} \mathcal{D}(u,v,w)_k(r)e^{ikx}.$$

We need to prove that

$$\|\mathcal{D}(u,v,w)\|_{X_\delta^{s+a,-1/2+\epsilon}} \lesssim \|u\|_{X_\delta^{s,1/2}}\|v\|_{X_\delta^{s,1/2}}\|w\|_{X_\delta^{s,1/2}},$$

for all sufficiently small $\epsilon > 0$. As usual, this follows by considering the $X^{s,b}$ norms instead of the restricted norms. By duality, it suffices to prove that

$$\Big|\sum_k \int_{\mathbb{R}} \widehat{\mathcal{D}}(\tau,k)\widehat{h}(-\tau,-k)d\tau\Big| = \Big|\int_{\mathbb{R}\times\mathbb{T}} \mathcal{D}(u,v,w)h\Big| \tag{4.20}$$

$$\lesssim \|u\|_{X^{s,1/2}}\|v\|_{X^{s,1/2}}\|w\|_{X^{s,1/2}}\|h\|_{X^{-s-a,\frac{1}{2}-\epsilon}}.$$

We note that

$$\widehat{\mathcal{D}}(\tau,k) = \frac{i}{6}\int_{\tau_1+\tau_2+\tau_3=\tau} \sum_{\substack{k_1+k_2+k_3=k \\ (k_2+k_3)(k_1+k_2)(k_1+k_3)\neq 0}} \frac{\widehat{u}(\tau_1,k_1)\widehat{v}(\tau_2,k_2)\widehat{w}(\tau_3,k_3)}{k_1}.$$

Let

$$f_1(\tau,k) = |\widehat{u}(\tau,k)||k|^s\langle\tau-k^3\rangle^{1/2},$$
$$f_2(\tau,k) = |\widehat{v}(\tau,k)||k|^s\langle\tau-k^3\rangle^{1/2},$$
$$f_3(\tau,k) = |\widehat{w}(\tau,k)||k|^s\langle\tau-k^3\rangle^{1/2},$$
$$f_4(\tau,k) = |\widehat{h}(\tau,k)||k|^{-s-a}\langle\tau-k^3\rangle^{\frac{1}{2}-\epsilon}.$$

Note that (4.20) follows from

$$\sum_{\substack{k_1+k_2+k_3+k_4=0 \\ (k_2+k_3)(k_1+k_2)(k_1+k_3)\neq 0}} \int_{\tau_1+\tau_2+\tau_3+\tau_4=0} \frac{|k_1k_2k_3|^{-s}|k_4|^{s+a}\prod_{i=1}^4 f_i(\tau_i,k_i)}{|k_1|\prod_{i=1}^4\langle\tau_i-k_i^3\rangle^{1/2-\epsilon}}$$

$$\lesssim \prod_{i=1}^4 \|f_i\|_2. \tag{4.21}$$

By Corollary 3.17, we have (for any $\epsilon > 0$)

$$\left\|\mathcal{F}^{-1}\left(\frac{f_i|k|^{-\epsilon}}{\langle\tau-k^3\rangle^{1/2+\epsilon}}\right)\right\|_{L^6(\mathbb{R}\times\mathbb{T})} \lesssim \|f_i\|_2. \tag{4.22}$$

Using $\tau_1+\tau_2+\tau_3+\tau_4 = 0$ and $k_1+k_2+k_3+k_4 = 0$, we have

$$\sum_{i=1}^4 \tau_i - k_i^3 = -k_1^3 - k_2^3 - k_3^3 - k_4^3 = 3(k_1+k_2)(k_1+k_3)(k_2+k_3).$$

Therefore

$$\max_{i=1,2,3,4} \langle \tau_i - k_i^3 \rangle \gtrsim |k_1 + k_2||k_1 + k_3||k_2 + k_3|.$$

We assume that

$$\langle \tau_1 - k_1^3 \rangle = \max_{i=1,2,3,4} \langle \tau_i - k_i^3 \rangle \gtrsim |k_1 + k_2||k_1 + k_3||k_2 + k_3|,$$

the other cases are similar. This implies that

$$\prod_{i=1}^{4} \langle \tau_i - k_i^3 \rangle^{\frac{1}{2}-\epsilon} \gtrsim (|k_1 + k_2||k_1 + k_3||k_2 + k_3|)^{\frac{1}{2}-7\epsilon} \prod_{i=2}^{4} \langle \tau_i - k_i^3 \rangle^{\frac{1}{2}+\epsilon}. \tag{4.23}$$

Now we claim that for $s > -\frac{1}{2}$, $a < \min(1, 2s + 1)$ and for ϵ sufficiently small

$$\frac{|k_1 k_2 k_3|^{-s} |k_4|^{s+a}}{|k_1| \left(|k_1 + k_2||k_1 + k_3||k_2 + k_3|\right)^{\frac{1}{2}-7\epsilon}} \lesssim |k_1 k_2 k_3 k_4|^{-\epsilon}. \tag{4.24}$$

Using (4.24) and (4.23) in (4.21) (and eliminating $|k_1|^{-\epsilon}$), we see that the left-hand side of (4.21) is

$$\lesssim \sum_{k_1+k_2+k_3+k_4=0} \int_{\tau_1+\tau_2+\tau_3+\tau_4=0} \frac{|k_2 k_3 k_4|^{-\epsilon} \prod_{i=1}^{4} f_i(\tau_i, k_i)}{\prod_{i=2}^{4} \langle \tau_i - k_i^3 \rangle^{\frac{1}{2}+\epsilon}}.$$

By Plancherel and the convolution structure, we can rewrite this as

$$\int_{\mathbb{R}\times\mathbb{T}} \widehat{f_1}(t,x) \prod_{i=2}^{4} \mathcal{F}^{-1}\left(\frac{f_i |k|^{-\epsilon}}{\langle \tau - k^3 \rangle^{\frac{1}{2}+\epsilon}} \right)(t,x) dt dx$$

$$\leq \|f_1\|_{L^2(\mathbb{R}\times\mathbb{T})} \prod_{i=2}^{4} \left\| \mathcal{F}^{-1}\left(\frac{f_i |k|^{-\epsilon}}{\langle \tau - k^3 \rangle^{\frac{1}{2}+\epsilon}} \right) \right\|_{L^6(\mathbb{R}\times\mathbb{T})} \lesssim \prod_{i=1}^{4} \|f_i\|_2.$$

In the first inequality, we used Hölder's inequality and Planchrel, and in the second one, we used (4.22).

It remains to prove the claim (4.24). First note that (since all factors are nonzero and $k_1 + k_2 + k_3 + k_4 = 0$)

$$|k_1|^3 + |k_2|^3 + |k_3|^3 \gtrsim |k_1 + k_2||k_1 + k_3||k_2 + k_3| \gtrsim |k_i|, \quad i = 1, 2, 3, 4. \tag{4.25}$$

By (4.25), and the fact that a, s are given in open intervals, (4.24) follows from

$$\frac{|k_1 k_2 k_3|^{-s} |k_4|^{s+a}}{|k_1| (|k_1 + k_2||k_1 + k_3||k_2 + k_3|)^{1/2}} \lesssim 1$$

with slightly different a and s satisfying the hypothesis of the proposition.

First consider the case $s > -\frac{1}{3}$, $a < \min(1, 2s + 1)$.

Without loss of generality, we can assume that $s + a \geq 0$. Let $M = \max(|k_1|, |k_2|, |k_3|)$. Using

$$|k_1||k_1 + k_2| \gtrsim |k_2| \text{ and } |k_1||k_1 + k_3||k_3 + k_2| \gtrsim |k_2|,$$

and by the symmetry of k_2, k_3, we have

$$|k_1| \left(|k_1 + k_2||k_1 + k_3||k_2 + k_3|\right)^{1/2} \gtrsim M.$$

Thus

$$\frac{|k_1 k_2 k_3|^{-s}|k_4|^{s+a}}{|k_1| \left(|k_1 + k_2||k_1 + k_3||k_2 + k_3|\right)^{1/2}} \lesssim \frac{|k_1 k_2 k_3|^{-s}|k_4|^{s+a}}{M}.$$

Since $|k_1 k_2 k_3|^{-s} \leq M^{-3s}$ for $s < 0$ and $|k_1 k_2 k_3|^{-s} \leq M^{-s}$ for $s \geq 0$, we have $|k_1 k_2 k_3|^{-s} \lesssim M^{-\min(s,3s)}$. Using this, the inequality $|k_4| \lesssim M$, and $0 \leq s + a < \min(3s + 1, s + 1)$, we bound the multiplier by

$$M^{-\min(s,3s)} M^{s+a-1} \lesssim 1.$$

Second consider the case $-\frac{1}{2} < s \leq -\frac{1}{3}$, $s+a < 3s+1 = s+\min(2s+1, 1) \leq 0$. Using

$$|k_4| = |k_1 + k_2 + k_3| \text{ and } |k_1 + k_2 + k_3||k_2 + k_3| \gtrsim |k_1|,$$

we have

$$\frac{|k_1 k_2 k_3|^{-s}|k_4|^{s+a}}{|k_1|(|k_1 + k_2||k_1 + k_3||k_2 + k_3|)^{\frac{1}{2}}} \lesssim \frac{|k_2 k_3|^{-s}}{|k_1|^{1-a}(|k_1 + k_2||k_1 + k_3|)^{1/2}|k_2 + k_3|^{\frac{1}{2}+s+a}}.$$

Now using $|k_1||k_1 + k_i| \gtrsim |k_i|$, we bound the multiplier by

$$\frac{|k_2 k_3|^{-s-\frac{1-a}{2}}}{(|k_1 + k_2||k_1 + k_3|)^{\frac{a}{2}}|k_2 + k_3|^{\frac{1}{2}+s+a}} \lesssim |k_2 k_3|^{\frac{a-(2s+1)}{2}} \lesssim 1.$$

In the last inequality, we estimated the denominator away since $\frac{1}{2} + s + a \geq 0$ and $a \geq 0$. This finishes the proof of (4.24), and the proposition.

4.2 High–low decomposition method

In this section, we present a method for obtaining global-in-time solutions evolving from infinite energy initial data. As we discussed before, local well-posedness in H^s together with an a priori bound coming from a conservation law in the same H^s level lead to global-in-time solutions. In this case, the local solutions can be iterated with a uniform time step. In the case of infinite energy solutions, Bourgain in [19] introduced a method for iterating the local solutions up to arbitrarily large time intervals by decomposing the data into low and high frequency components. The evolution of the low frequency component is global because of conservation laws. For the difference equation in the local time interval, a smoothing bound as in Section 4.1 enables one to write the solution as the linear evolution plus a small nonlinear term. Since the linear evolution preserves any H^s norm, one can continue the iteration indefinitely covering any time interval.

We demonstrate this method for the KdV equation with a smooth and mean-zero space-time potential V

$$\begin{cases} u_t + u_{xxx} + \left(Vu + \frac{u^2}{2}\right)_x = 0, & x \in \mathbb{T}, \quad t \in \mathbb{R}, \\ u(0, \cdot) = g(\cdot) \in H^s(\mathbb{T}). \end{cases} \tag{4.26}$$

The local wellposedness in $H^s(\mathbb{T})$, $s > -\frac{1}{2}$, was presented in Section 3.3.3 for mean-zero data;see Theorem 3.21. In this section, we need a slightly different version that can be easily deduced from the proof of Theorem 3.21:

Theorem 4.8 *Let V be a mean-zero space-time potential. Given mean-zero $g \in H^s$ and $\delta > 0$ satisfying*

$$\delta \lesssim [\|g\|_{H^s} + \|V\|_{Y^s_\delta}]^{-6-},$$

there exists a unique solution u to (4.26) *in Y^s_δ, and it satisfies*

$$\|u\|_{Y^s_\delta} \lesssim \|g\|_{H^s}.$$

We further have the growth bound (see the calculations before (3.33))

$$\|u(t_2)\|_{L^2} \leq \|u(t_1)\|_{L^2} \exp\left(\frac{1}{2}\int_{t_1}^{t_2} \|V_x(\tau)\|_{L^\infty} d\tau\right). \tag{4.27}$$

This gives global wellposedness in the H^s level, $s \geq 0$. The following theorem establishes global-in-time solutions below the L^2 level:

Theorem 4.9 *Let $-\frac{1}{21} < s < 0$. Let $V \in C^2_t H^2_x$ be a mean-zero space-time potential. The KdV equation* (4.26) *is globally wellposed for any mean-zero $g \in H^s(\mathbb{T})$. Moreover*

$$u(t) - W_t g \in C^0_t H^{s+a}_x(\mathbb{R} \times \mathbb{T})$$

for $a < 2s + 1$, where $W_t = e^{-t\partial_{xxx}}$.

Proof Fix large N to be determined later, and write $g = \phi_0 + \psi_0$, where $\phi_0 = P_N(g)$, with $\widehat{P_N(g)} = \chi_{[-N,N]}\widehat{g}$. We have

$$\|\phi_0\|_{L^2} \leq N^{-s}\|g\|_{H^s}.$$

We decompose the solution as $u = v + w$, where v solves the equation

$$\begin{cases} v_t + v_{xxx} + \left(Vv + \frac{v^2}{2}\right)_x = 0, & x \in \mathbb{T}, \quad t \in \mathbb{R}, \\ v(0,x) = \phi_0(x), \end{cases} \tag{4.28}$$

and w solves the equation

$$\begin{cases} w_t + w_{xxx} + \left(\Lambda w + \frac{w^2}{2}\right)_x = 0, & x \in \mathbb{T}, \quad t \in \mathbb{R}, \\ w(0,x) = \psi_0(x), \end{cases} \tag{4.29}$$

where $\Lambda = V + v$.

It is important to note that although

$$\|\psi_0\|_{H^s} \leq \|g\|_{H^s},$$

for any $-\frac{1}{2} < s_0 < s$ we have the much better bound

$$\|\psi_0\|_{H^{s_0}} \leq N^{s_0 - s}\|g\|_{H^s}.$$

By Theorem 4.8 the equation (4.28) has a unique solution in Y^0_δ with

$$\delta \lesssim \left[\|\phi_0\|_{L^2} + \|g\|_{H^s} + \|V\|_{Y^0_\delta}\right]^{-6-}$$

satisfying

$$\|v\|_{X^{0,1/2}_\delta} \leq \|v\|_{Y^0_\delta} \lesssim \|\phi_0\|_{L^2}.$$

Noting that

$$\|\Lambda\|_{Y^{s_0}_\delta} \leq \|\Lambda\|_{Y^0_\delta} \leq \|V\|_{Y^0_\delta} + \|v\|_{Y^0_\delta} \lesssim \|V\|_{Y^0_\delta} + \|\phi_0\|_{L^2},$$

and

$$\|\psi_0\|_{H^{s_0}} \leq N^{s_0 - s}\|g\|_{H^s} \leq \|g\|_{H^s}$$

we see that Theorem 4.8 also applies to the equation (4.29) with the same δ and we have

$$\|w\|_{H^{s_0}} \lesssim \|w\|_{Y^{s_0}_\delta} \lesssim \|\psi_0\|_{H^{s_0}} \leq N^{s_0 - s}\|g\|_{H^s}.$$

Later on we will choose δ smaller to guarantee that it works for every step of the iteration. From now on all implicit constants depend on $\|g\|_{H^s}$. Using the differentiation by parts calculation from Section 4.1.2 equation (4.15) (with $f = \gamma = 0$ and $V = \Lambda$), we can rewrite the system (4.29) as

$$\begin{aligned} w(t) - W_t\psi_0 = &-\mathcal{B}(2\Lambda + w, w)(t) + W_t\mathcal{B}(2\Lambda(0) + \psi_0, \psi_0) \\ &+ 2\int_0^t W_{t-r}\mathcal{B}\left(W_r\partial_r(W_{-r}\Lambda), w\right)(r)dr \\ &+ \int_0^t W_{t-r}\tilde{\rho}(r)dr + \int_0^t W_{t-r}\mathcal{D}(\Lambda + w, 2\Lambda + w, w)(r)dr, \end{aligned} \tag{4.30}$$

where

$$\mathcal{B}(f,g)_k = -\frac{1}{6}\sum_{k_1+k_2=k}\frac{f_{k_1}g_{k_2}}{k_1k_2},$$

$$\tilde{\rho}_k = \frac{i}{3}\Lambda_k\sum_{|j|\neq|k|}\frac{\Lambda_j\overline{w_j}}{j} - \frac{i}{6}\frac{(\overline{\Lambda_k}+\overline{w_k})(2\Lambda_k+w_k)w_k}{k}$$

$$\mathcal{D}(f,g,h)_k = \frac{i}{6}\sum_{\substack{k_1+k_2+k_3=k\\(k_2+k_3)(k_1+k_2)(k_1+k_3)\neq 0}}\frac{f_{k_1}g_{k_2}h_{k_3}}{k_1}.$$

Using this representation, we write

$$w(t) = W_t\psi_0 + z(t).$$

We will prove that $\|z\|_{L^2}$ remains small in the interval $[0,\delta]$. Using Lemma 4.5, provided that $s_0 > -1/3$, we have

$$\|\mathcal{B}(2\Lambda + w, w)\|_{L^2} \lesssim (\|\Lambda\|_{L^2} + \|w\|_{H^{s_0}})\|w\|_{H^{s_0}} \lesssim (\|V\|_{L^\infty_t L^2_x} + \|\phi_0\|_{L^2})N^{s_0-s},$$

and

$$\left\|\int_0^t W_{t-r}\tilde{\rho}(r)dr\right\|_{L^2}$$
$$\lesssim \int_0^t (\|\Lambda(r)\|_{L^2} + \|w(r)\|_{H^{s_0}})^2\,\|w(r)\|_{H^{s_0}}dr$$
$$\lesssim \delta\left(\|V\|_{L^\infty_t L^2_x} + \|\phi_0\|_{L^2}\right)^2 N^{s_0-s}.$$

To estimate the L^2 norm of the third summand in (4.30), first note that

$$\mathcal{B}(W_r\partial_r(W_{-r}\Lambda), w) = \mathcal{B}(W_r\partial_r(W_{-r}V), w) + \mathcal{B}(W_r\partial_r(W_{-r}v), w). \quad (4.31)$$

Contribution of the first summand in (4.31) to the L^2 norm of the third summand in (4.30) can be estimated by

$$\lesssim \int_0^t \|(-L+\partial_r)V\|_{L^2}\|w(r)\|_{H^{s_0}}dr \lesssim \delta N^{s_0-s}\|(-L+\partial_t)V\|_{L^\infty_t L^2_x}.$$

To estimate the contribution of the second summand, note that

$$W_r\partial_r(W_{-r}v) = -Lv + v_r = -\left(Vv + \frac{v^2}{2}\right)_x.$$

Using this, we have

$$\|\mathcal{B}(W_r\partial_r(W_{-r}v), w)\|_{L^2} \lesssim \left\| \sum_{k_1+k_2=k} \mathcal{F}\left(Vv + v^2/2\right)(k_1)\frac{w_{k_2}}{k_2}\right\|_{\ell^2}$$
$$\lesssim \left\|Vv + v^2/2\right\|_{L^2} \left\|\frac{w_{k_2}}{k_2}\right\|_{\ell^1} \lesssim \left\|Vv + v^2/2\right\|_{L^2} \|w\|_{H^{s_0}}$$
$$\leq \left(\|V\|_{L^4}\|v\|_{L^4} + \|v\|_{L^4}^2\right)\|w\|_{H^{s_0}}.$$

Therefore, applying Hölder's inequality in the time integral, the contribution to the L^2 norm of the third summand in (4.30) is

$$\lesssim \left(\delta^{1/2}\|v\|^2_{L^4_{t\in[0,\delta]}L^4_x} + \delta^{3/4}\|v\|_{L^4_{t\in[0,\delta]}L^4_x}\|V\|_{L^\infty_t L^4_x}\right)\|w\|_{L^\infty_{t\in[0,\delta]}H^{s_0}}.$$

Utilizing Theorem 3.18 and then Lemma 3.11, we bound this by

$$\lesssim \left(\delta^{1/2}\|v\|^2_{X^{0,1/3}_\delta} + \delta^{3/4}\|v\|_{X^{0,1/3}_\delta}\|V\|_{L^\infty_t L^4_x}\right)\|w\|_{L^\infty_{t\in[0,\delta]}H^{s_0}}$$
$$\lesssim \left(\delta^{5/6-}\|\phi_0\|^2_{L^2} + \delta^{11/12-}\|\phi_0\|_{L^2}\|V\|_{L^\infty_t L^4_x}\right)N^{s_0-s}.$$

Finally, $s_0 > -1/3$ Proposition 4.7 implies that

$$\left\|\int_0^t W_{t-r}\mathcal{D}(\Lambda + w, 2\Lambda + w, w)(r)dr\right\|_{L^2}$$
$$\lesssim \left\|\int_0^t W_{t-r}\mathcal{D}(\Lambda + w, 2\Lambda + w, w)(r)dr\right\|_{X^{0,1/2+}_\delta}$$
$$\lesssim \|\mathcal{D}(\Lambda + w, 2\Lambda + w, w)\|_{X^{0,-1/2+}_\delta}$$
$$\lesssim \left(\|\Lambda\|_{X^{s_0,1/2}_\delta} + \|w\|_{X^{s_0,1/2}_\delta}\right)^2 \|w\|_{X^{s_0,1/2}_\delta}$$
$$\lesssim \left(\|V\|_{C^2_t H^2_x} + \|\phi_0\|_{L^2}\right)^2 N^{s_0-s}.$$

Using the estimates above, we obtain on $[0, \delta]$

$$\|z\|_{L^2} \lesssim \|\phi_0\|^2_{L^2} N^{s_0-s}.$$

We note that when there is no potential the estimates above would give a better bound for $\|z\|_{L^2}$.

Now fix $t_1 \in [0, \delta]$ and write $u(t_1) = \phi_1 + \psi_1$, where

$$\phi_1 = v(t_1) + z(t_1), \quad \text{and } \psi_1 = W_{t_1}\psi_0.$$

Note that ψ_1 has the same properties as ψ_0, more explicitly

$$\|\psi_1\|_{H^{s_0}} = \|\psi_0\|_{H^{s_0}} \leq N^{s_0-s}\|g\|_{H^s}, \quad \|\psi_1\|_{H^s} \leq \|g\|_{H^s},$$

and

$$\begin{aligned}\|\phi_1\|_{L^2} &\le \|v(t_1)\|_{L^2} + \|z(t_1)\|_{L^2} \\ &\le \|v(0)\|_{L^2} \exp\left(\frac{1}{2}\int_0^{t_1} \|V_x(\tau)\|_{L^\infty} d\tau\right) + C\|\phi_0\|_{L^2}^2 N^{s_0-s} \\ &= \|\phi_0\|_{L^2} \exp\left(\frac{1}{2}\int_0^{t_1} \|V_x(\tau)\|_{L^\infty} d\tau\right) + C\|\phi_0\|_{L^2}^2 N^{s_0-s}.\end{aligned}$$

Similarly at time $t_k = k\delta$, we write

$$u(t_k) = \phi_k + \psi_k,$$

where $\psi_k = W_{k\delta}\psi_0$ has the same properties like ψ_0, and

$$\|\phi_k\|_{L^2} \le \|\phi_{k-1}\|_{L^2} \exp\left(\frac{1}{2}\int_{t_{k-1}}^{t_k} \|V_x(\tau)\|_{L^\infty} d\tau\right) + C\|\phi_{k-1}\|_{L^2}^2 N^{s_0-s}.$$

Fix T large. Let

$$A_T := \exp\left(\frac{1}{2}\int_0^T \|V_x\|_{L^\infty} d\tau\right).$$

We will choose N large enough and

$$\delta \sim (N^{-s}A_T)^{-6-}. \tag{4.32}$$

We claim that if N is large enough

$$\|\phi_k\|_{L^2} \le 2A_T\|\phi_0\|_{L^2} \text{ for } k = 1, 2, ..., T/\delta.$$

Indeed, using the notation

$$a_k = \exp\left(\frac{1}{2}\int_{t_{k-1}}^{t_k} \|V_x(\tau)\|_{L^\infty} d\tau\right),$$

we inductively have

$$\begin{aligned}\|\phi_k\|_{L^2} &\le a_k\|\phi_{k-1}\|_{L^2} + C\|\phi_{k-1}\|_{L^2}^2 N^{s_0-s} \\ &\le a_k\|\phi_{k-1}\|_{L^2} + 4CA_T^2\|\phi_0\|_{L^2}^2 N^{s_0-s} \\ &\le a_k a_{k-1}\|\phi_{k-2}\|_{L^2} + (a_k + 1)4CA_T^2\|\phi_0\|_{L^2}^2 N^{s_0-s} \\ &\le a_k a_{k-1} a_{k-2}\|\phi_{k-3}\|_{L^2} + (a_{k-1}a_k + a_k + 1)4CA_T^2\|\phi_0\|_{L^2}^2 N^{s_0-s}.\end{aligned}$$

Iterating this process, we obtain (for $k \le K = T/\delta$)

$$\begin{aligned}
\|\phi_k\|_{L^2} &\le \left[\prod_{j=1}^{k} a_j\right] \|\phi_0\|_{L^2} + \left[1 + \sum_{i=0}^{k-2} \prod_{j=k-i}^{k} a_j\right] 4CA_T^2 \|\phi_0\|_{L^2}^2 N^{s_0-s} \\
&\le A_T \|\phi_0\|_{L^2} + 4CkA_T^3 \|\phi_0\|_{L^2}^2 N^{s_0-s} \\
&\le A_T \|\phi_0\|_{L^2} \left[1 + 4CA_T^2 T\delta^{-1} \|\phi_0\|_{L^2} N^{s_0-s}\right].
\end{aligned}$$

Using (4.32), we bound this by

$$\le A_T \|\phi_0\|_{L^2} \left[1 + \widetilde{C} T A_T^{8+} N^{s_0-7s-}\right] \le 2A_T \|\phi_0\|_{L^2},$$

provided that $s > \frac{s_0}{7} > -\frac{1}{21}$, and

$$N \approx \left(TA_T^{8+}\right)^{-\frac{1}{s_0-7s-}} \approx \left(TA_T^{8+}\right)^{\frac{3}{1+21s}+}.$$

This yields the global wellposedness result as stated. Also note that for any $0 \ge s > -\frac{1}{21}$ we have

$$\begin{aligned}
\|u(T) - W_T g\|_{L^2} &= \|u(T) - W_T \psi_0 - W_T \phi_0\|_{L^2} \\
&= \|u(T) - \psi_K - W_T \phi_0\|_{L^2} = \|\phi_K - W_T \phi_0\|_{L^2} \\
&\lesssim A_T \|\phi_0\|_{L^2} + \|\phi_0\|_{L^2} \lesssim A_T N^{-s} \lesssim A_T (TA_T^{8+})^{-\frac{3s}{1+21s}}.
\end{aligned}$$

Moreover

$$\|u(T)\|_{H^s} \le \|u(T) - W_T g\|_{L^2} + \|W_T g\|_{H^s} \lesssim 1 + A_T (TA_T^{8+})^{-\frac{3s}{1+21s}}.$$

Finally, using this a priori bound in Theorem 4.3 yields the smoothing claim of Theorem 4.9. □

We note that the hypothesis on s in Theorem 4.9 can be weakened in the case $V = 0$; see Exercise 4.6.

For applications of the high–low decomposition method to other equations, see, e.g., Colliander–Staffilani–Takaoka [46] and Demirbaş–Erdoğan–Tzirakis [51]. For an application to a dispersive system, see Pecher [122].

4.3 The *I*-method for the quintic NLS equation on the torus

In this section, we present another globalizing technique, the *I*-method, to obtain global wellposedness for infinite energy initial data. The method was introduced by Colliander–Keel–Staffilani–Takaoka–Tao [41, 42]. It was inspired by a paper of Keel–Tao [86], where the authors proved the global wellposedness below the energy for wave maps on $\mathbb{R}$.

Consider the quintic defocusing Schrödinger equation on the torus:

$$\begin{cases} iu_t + u_{xx} - |u|^4 u = 0, & t \in \mathbb{R}, \ x \in \mathbb{T}, \\ u(0, \cdot) = g(\cdot) \in H^s(\mathbb{T}). \end{cases} \tag{4.33}$$

The problem is locally wellposed for any $s > 0$ as it was proved by Bourgain [15, 20]; also see Theorem 3.31. For $s > \frac{1}{2}$, the local wellposedness follows easily from the Sobolev embedding theorem (see Exercise 3.5), which suffices for our purposes. It is also known that the data-to-solution map cannot be C^5 if $g \in L^2(\mathbb{T})$, see Kishimoto [95]. For $s < 0$, Christ–Colliander–Tao [39] obtained a stronger illposedness result. More precisely, they showed that the data-to-solution map is not continuous from $H^s(\mathbb{T})$ to the space of periodic distributions.

We note that the smooth solutions of the equation (4.33) satisfy the following conservation laws:

The conservation of mass

$$\|u(t)\|_{L^2(\mathbb{T})} = \|g\|_{L^2(\mathbb{T})},$$

and the conservation of energy

$$\begin{aligned} E(u)(t) &= \frac{1}{2}\int_{\mathbb{T}} |u_x(t,x)|^2\, dx + \frac{1}{6}\int_{\mathbb{T}} |u(t,x)|^6\, dx \\ &= E(u)(0) = \frac{1}{2}\int_{\mathbb{T}} |g_x(x)|^2\, dx + \frac{1}{6}\int_{T} |g(x)|^6\, dx. \end{aligned}$$

The two conserved quantities imply the bound

$$\sup_{t\in\mathbb{R}} \|u(t)\|_{H^1(\mathbb{T})} \le C_{\|g\|_{H^1}},$$

which can be iterated to give global wellposedness for H^1 data. In this section, we use the I-method to prove global wellposedness for solutions that evolve from initial data with infinite energy, H^s, $s < 1$. The result we present is not the best one. More precisely, Bourgain in [21] combined the I-method with normal form techniques and proved global wellposedness in $H^{\frac{1}{2}-}(\mathbb{T})$. A more recent result proves global wellposedness in $H^{\frac{2}{5}+}$; see Li–Wu–Xu [103].

In Section 4.2, we already presented Bourgain's high–low decomposition method, [19]. The I-method is a refinement of Bourgain's method, and it is based on the almost conservation of a certain modified energy functional. The idea is to replace the conserved quantity $E(u)$ with an "almost conserved" variant $E(Iu)$, where I is a smoothing operator of order $1 - s$. This operator behaves like the identity operator for low frequencies and like a fractional integral operator for high frequencies.

It is clear that an a priori estimate of the form

$$\sup_{t\in\mathbb{R}} \|u(t)\|_{H^s} \leq C_{\|g\|_{H^s}}$$

would imply global wellposedness in H^s. Absent a conservation law at this level, it is very hard to obtain such a bound. However, one can relax the boundedness of the Sobolev norm, with a growth bound and still be able to prove the global wellposedness. For example, if we know that the solutions are bounded only polynomially in time

$$\sup_{t\in[0,T]} \|u(t)\|_{H^s} \leq C_{\|g\|_{H^s}}(1+T)^p,$$

then global wellposedness would follow.

To this end, we introduce a radial, C^∞, and monotone multiplier m, taking values in $[0,1]$, and

$$m(k) := \begin{cases} 1, & |k| < N \\ \left(\frac{|k|}{N}\right)^{s-1}, & |k| > 2N. \end{cases} \tag{4.34}$$

Here N is a large parameter to be determined later. We define an operator I by

$$\widehat{Iu}(k) = m(k)\widehat{u}(k), \quad k \in \mathbb{Z}. \tag{4.35}$$

This operator is smoothing of order $1-s$, and the following estimate holds; see Exercise 4.9,

$$\|u\|_{H^s} \lesssim \|Iu\|_{H^1} \lesssim N^{1-s}\|u\|_{H^s}. \tag{4.36}$$

Thus, the I operator acts as the identity for low frequencies while it maps H^s solutions to H^1. Also note that it commutes with all derivatives. With the aid of this operator, we define the modified energy functional

$$E(Iu)(t) = \frac{1}{2}\int_{\mathbb{T}} |Iu_x(t)|^2 dx + \frac{1}{6}\int_{\mathbb{T}} |Iu(t)|^6 dx.$$

Moreover, since the equation is defocusing and the multiplier $m(k) \leq 1$, we have that

$$\|Iu\|_{H^1}^2 = \|Iu_x\|_{L^2}^2 + \|Iu\|_{L^2}^2 \leq 2E(Iu) + \|u\|_{L^2}^2 = 2E(Iu) + \|g\|_{L^2}^2.$$

Therefore

$$\|u\|_{H^s}^2 \lesssim \|Iu\|_{H^1}^2 \lesssim E(Iu) + \|g\|_{L^2}^2. \tag{4.37}$$

This implies that we can control the H^s norm of the solution if we can control the quantity $E(Iu)$. Of course, this quantity is not conserved in time since Iu

does not satisfy the equation. Instead, by applying the I operator to the equation (4.33), it formally satisfies the following

$$i(Iu)_t + (Iu)_{xx} = I(|u|^4 u) = F(Iu) + [I, F](u), \tag{4.38}$$

where $F(u) = |u|^4 u$, and

$$[I, F](u) = I(F(u)) - F(Iu)$$

is the commutator of I and the nonlinearity $F(u)$. The method will be successful if this commutator remains small as the problem evolves. The bound on the commutator leads to an almost conservation law for the new energy functional $E(Iu)$. The key is to prove that on the local wellposedness interval the increment of the modified energy $E(Iu)$ decays with respect to a large parameter N.

Using the fundamental theorem of calculus and equation (4.38) for Iu, we obtain that

$$\begin{aligned} E(Iu)(T) - E(Iu)(0) &= \int_0^T \frac{\partial}{\partial t} E(Iu(t))\, dt \\ &= \Re \int_0^T \int_{\mathbb{T}} \overline{Iu_t}\, (-Iu_{xx} + F(Iu))\, dxdt \\ &= \Re \int_0^T \int_{\mathbb{T}} \overline{Iu_t}\, (F(Iu) - IF(u))\, dxdt, \end{aligned}$$

where in the last step we used the fact that

$$-Iu_{xx} = iIu_t - IF(u).$$

Using again the same identity in the form

$$\overline{Iu_t} = -i\overline{Iu_{xx}} + i\overline{IF(u)},$$

we obtain that

$$\begin{aligned} E(Iu)(T) - E(Iu)(0) = -\Im \int_0^T \int_{\mathbb{T}} \overline{Iu_{xx}}\, (F(Iu) - IF(u))\, dxdt\, + \\ \Im \int_0^T \int_{\mathbb{T}} \overline{IF(u)}\, (F(Iu) - IF(u))\, dxdt. \end{aligned} \tag{4.39}$$

Notice that this implies the conservation of energy taking I to be the identity operator. In our case, the nonlinearity is algebraic and one can write the commutator between the I operator and the nonlinearity explicitly using the Fourier transform, and control it by multilinear estimates. This analysis can be carried out in the setting of $X^{s,b}$ spaces, where one can use a variety of linear and bilinear Strichartz estimates both on $\mathbb{R}$ and $\mathbb{T}$.

The first step is to prove a modified local wellposedness theory for the Iu equation (4.38) at the H^1 level. Recall that the $X^{s,b}$ space is defined by the norm

$$\|u\|_{X^{s,b}} = \left\| \langle k \rangle^s \left\langle \tau + k^2 \right\rangle^b \widehat{u}(\tau, k) \right\|_{L^2_\tau \ell^2_k}.$$

Proposition 4.10 *Fix $\frac{1}{2} < s < 1$, and consider the initial value problem*

$$\begin{cases} iIu_t + Iu_{xx} = I\left(|u|^4 u\right), & x \in \mathbb{T}, \quad t \in [-T, T], \\ Iu(0, x) = Ig(x) \in H^1(\mathbb{T}). \end{cases} \tag{4.40}$$

Then equation (4.40) *is locally wellposed in $H^1(\mathbb{T})$ on an interval $[-\delta, \delta]$ with $\delta \sim \|Ig\|_{H^1(\mathbb{T})}^{-4-}$ and satisfies the bound*

$$\|Iu\|_{X_\delta^{1,\frac{1}{2}+}} \lesssim \|Ig\|_{H^1(\mathbb{T})}.$$

Proof Recall Duhamel's formula

$$Iu(t, x) = e^{it\partial_{xx}} Ig(x) - i \int_0^t e^{i(t-s)\partial_{xx}} I\left(|u|^4 u\right)(s, x) ds.$$

Also note that using Lemma 3.12, we have

$$\left\| \int_0^t e^{i(t-s)\partial_{xx}} I(|u|^4 u)(s, x) ds \right\|_{X_\delta^{1,b}} \lesssim \delta^{1+b'-b} \left\| I(|u|^4 u) \right\|_{X_\delta^{1,b'}}.$$

Therefore, for $b = \frac{1}{2}+$ and $b' = 0$, we have

$$\left\| \int_0^t e^{i(t-s)\partial_{xx}} I(|u|^4 u)(s, x) ds \right\|_{X_\delta^{1,\frac{1}{2}+}} \lesssim \delta^{\frac{1}{2}-} \left\| I(|u|^4 u) \right\|_{X_\delta^{1,0}}.$$

Since in addition for $\delta \leq 1$

$$\left\| e^{it\partial_{xx}} Ig(x) \right\|_{X_\delta^{1,\frac{1}{2}+}} \lesssim \|Ig\|_{H^1},$$

we obtain that

$$\|Iu\|_{X_\delta^{1,\frac{1}{2}+}} \lesssim \|Ig\|_{H^1} + \delta^{\frac{1}{2}-} \left\| I\left(|u|^4 u\right) \right\|_{X_\delta^{1,0}}.$$

Now we claim that

$$\|I(|u|^4 u)\|_{X_\delta^{1,0}} \lesssim \delta^{\frac{1}{2}-} \|Iu\|^5_{X_\delta^{1,\frac{1}{2}+}}. \tag{4.41}$$

Using this claim, we have

$$\|Iu\|_{X_\delta^{1,\frac{1}{2}+}} \lesssim \|Ig\|_{H^1} + \delta^{1-} \|Iu\|^5_{X_\delta^{1,\frac{1}{2}+}},$$

and using standard arguments one completes the proof. Note that to close the contraction argument, we need to take $\delta \lesssim \|Ig\|_{H^1}^{-4-}$.

It remains to show (4.41). By L^2 duality and ignoring the δ dependence, we have

$$\left\| I(|u|^4 u) \right\|_{X^{1,0}} = \sup_{\|v\|_{X^{-1,0}}=1} \left| \left\langle I(|u|^4 u), v \right\rangle_{L^2(\mathbb{T}\times\mathbb{R})} \right|.$$

Using Plancherel, we can write the complex conjugate of the inner product above as

$$\int\int I(|u|^4 u)\bar{v}\, dxdt =$$
$$\sum_{\sum_{i=1}^6 k_i=0} \int_{\sum_{i=1}^6 \tau_i=0} m(k_6)\widehat{u}(\tau_1,k_1)\widehat{\bar{u}}(\tau_2,k_2)\widehat{u}(\tau_3,k_3)\widehat{\bar{u}}(\tau_4,k_4)\widehat{u}(\tau_5,k_5)\widehat{\bar{v}}(\tau_6,k_6).$$

Now set

$$f_1(\tau_1,k_1) = |\widehat{u}(\tau_1,k_1)| m(k_1)\langle k_1\rangle\langle \tau_1 + k_1^2\rangle^{\frac{1}{2}+},$$

$$f_2(\tau_2,k_2) = |\widehat{u}(-\tau_2,-k_2)| m(k_2)\langle k_2\rangle\langle \tau_2 - k_2^2\rangle^{\frac{1}{2}+},$$

$$f_3(\tau_3,k_3) = |\widehat{u}(\tau_3,k_3)| m(k_3)\langle k_3\rangle\langle \tau_3 + k_3^2\rangle^{\frac{1}{2}+},$$

$$f_4(\tau_4,k_4) = |\widehat{u}(-\tau_4,-k_4)| m(k_4)\langle k_4\rangle\langle \tau_4 - k_4^2\rangle^{\frac{1}{2}+},$$

$$f_5(\tau_5,k_5) = |\widehat{u}(\tau_5,k_5)| m(k_5)\langle k_5\rangle\langle \tau_5 + k_5^2\rangle^{\frac{1}{2}+},$$

$$f_6(\tau_6,k_6) = |\widehat{v}(\tau_6,k_6)| \langle k_6\rangle^{-1}.$$

We need to show for nonnegative L^2 functions f_j that

$$\sum_{\sum_{j=1}^6 k_j=0} \int_{\sum_{j=1}^6 \tau_j=0} \frac{m(k_6)\langle k_6\rangle}{\prod_{j=1}^5 m(k_j)\langle k_j\rangle\langle \tau_j + (-1)^{j-1}k_j^2\rangle^{\frac{1}{2}+}} \prod_{j=1}^6 f_j(\tau_j,k_j)$$
$$\lesssim \prod_{j=1}^6 \|f_j\|_{L^2_\tau l^2_k}.$$

By Exercise 4.10, on the hyperplane $\sum_{j=1}^6 k_j = 0$ we have that

$$\frac{m(k_6)\langle k_6\rangle^{1-s}}{\prod_{j=1}^5 m(k_j)\langle k_j\rangle^{1-s}} \lesssim 1.$$

Therefore, it remains to show that (also see Exercise 4.11)

$$\sum_{\sum_{j=1}^{6} k_j=0} \int_{\sum_{j=1}^{6} \tau_j=0} \frac{\langle k_6\rangle^s}{\prod_{j=1}^{5}\langle k_j\rangle^s \langle \tau_j+(-1)^{j-1}k_j^2\rangle^{\frac{1}{2}+}} \prod_{j=1}^{6} f_j(\tau_j,k_j)$$

$$\lesssim \prod_{j=1}^{6} \|f_j\|_{L^2_\tau \ell^2_k}. \quad (4.42)$$

Noting that on the hyperplane $\sum_{j=1}^{6} k_j = 0$, we have that

$$\langle k_6\rangle^s \lesssim \sum_{j=1}^{5} \langle k_j\rangle^s,$$

and by symmetry it is enough to prove

$$\sum_{\sum_{j=1}^{6} k_j=0} \int_{\sum_{j=1}^{6} \tau_j=0} \frac{\langle k_1\rangle^s}{\prod_{j=1}^{5}\langle k_j\rangle^s \langle \tau_j+(-1)^{j-1}k_j^2\rangle^{\frac{1}{2}+}} \prod_{j=1}^{6} f_j(\tau_j,k_j) \lesssim \prod_{j=1}^{6} \|f_j\|_{L^2_\tau \ell^2_k}.$$

As before, by duality, the convolution structure, and Plancherel, this is equivalent to proving that

$$\|(J^s h_1)h_2h_3h_4h_5\|_{L^2_t L^2_x} = \|(J^s h_1)h_2h_3h_4h_5\|_{X^{0,0}} \lesssim \prod_{j=1}^{5} \|h_j\|_{X^{s,\frac{1}{2}+}},$$

where

$$h_1 = \mathcal{F}^{-1}\left(\langle k_1\rangle^{-s}\langle \tau_1+k_1^2\rangle^{-\frac{1}{2}-} f_1(\tau_1,k_1)\right),$$

and other h_js are defined similarly. Using the continuous embeddings

$$X^{\frac{1}{2}+,\frac{1}{2}+} \subset C_t^0 H_x^{\frac{1}{2}+} \subset C_t^0 C_x^0,$$

we have

$$\|u\|_{L_t^\infty L_x^\infty} \lesssim \|u\|_{X^{\frac{1}{2}+,\frac{1}{2}+}}.$$

This estimate, along with Hölder's inequality, yields that

$$\|(J^s h_1)h_2h_3h_4h_5\|_{L^2_t L^2_x} \le \|J^s h_1\|_{L^2_t L^2_x} \prod_{j=2}^{5} \|h_j\|_{L^\infty L^\infty}$$

$$\lesssim \|h_1\|_{X^{s,0}} \prod_{j=2}^{5} \|h_j\|_{X^{s,\frac{1}{2}+}} \lesssim \prod_{j=1}^{5} \|h_j\|_{X^{s,\frac{1}{2}+}},$$

for any $s > \frac{1}{2}$. As usual, we ignored the δ dependence in this proof. The additional gain in δ follows from Lemma 3.11 in the last inequality above. The proof is complete. □

Having a well-defined solution Iu on the local interval $[0, \delta]$ given by Proposition 4.10, we can now control the increment in the modified energy $E(Iu)$ in the local interval.

Proposition 4.11 *Let Iu solve the initial value problem* (4.40) *on* $[0, \delta]$; *with δ given by Proposition 4.10, we have that*

$$\left|E(Iu)(\delta) - E(Iu)(0)\right| \lesssim \delta^{\frac{3}{4}-} N^{-1+} \|Iu\|^6_{X_\delta^{1,\frac{1}{2}+}} + \delta N^{-2+} \|Iu\|^{10}_{X_\delta^{1,\frac{1}{2}+}}.$$

Remark 4.12 Using Proposition 4.10 and (4.36), we have

$$\|Iu\|_{X_\delta^{1,\frac{1}{2}+}} \lesssim \|Ig\|_{H^1(\mathbb{T})} \lesssim N^{1-s}.$$

Therefore, an easy calculation shows that when $s > \frac{3}{4}$

$$N^{-2+} \|Iu\|^{10}_{X_\delta^{1,\frac{1}{2}+}} \lesssim N^{-1+} \|Iu\|^6_{X_\delta^{1,\frac{1}{2}+}}.$$

Using this in Proposition 4.11, we obtain the bound

$$|E(Iu)(\delta) - E(Iu)(0)| \lesssim \delta^{\frac{3}{4}-} N^{-1+} \|Iu\|^6_{X_\delta^{1,\frac{1}{2}+}}. \tag{4.43}$$

We momentarily assume the validity of Proposition 4.11, and use it to prove our main theorem.

Theorem 4.13 *Fix $s > \frac{4}{5}$. Let u be a solution to* (4.33) *with initial data $g \in H^s(\mathbb{T})$. Then for any $T > 0$, we have that*

$$\sup_{0 \le t \le T} \|u(t)\|_{H^s} \lesssim C_{(\|g\|_{H^s}, T)}.$$

In particular, the defocusing quintic NLS equation on the torus is globally wellposed in $H^s(\mathbb{T})$, $s > \frac{7}{8}$.

Proof Recall from (4.37) that

$$\|u\|^2_{H^s} \lesssim \|Iu\|^2_{H^1} \le \|g\|^2_{L^2} + E(Iu).$$

It remains to bound $E(Iu)$ for arbitrarily large times. Recall that

$$E(Ig)(t) = \frac{1}{2} \int_{\mathbb{T}} |Ig_x|^2 dx + \frac{1}{6} \int |Ig|^6 dx.$$

To bound the L^6 norm, we use the Gagliardo–Nirenberg inequality and we obtain that

$$E(Ig) \lesssim \|Ig_x\|_{L^2}^2 + \|Ig_x\|_{L^2}^2 \|Ig\|_{L^2}^4 \lesssim \|Ig\|_{H^1}^2 \left(1 + \|Ig\|_{L^2}^4\right)$$
$$\lesssim \|Ig\|_{H^1}^2 \left(1 + \|g\|_{L^2}^4\right) \lesssim C_{\|g\|_{H^s}} N^{2(1-s)}. \quad (4.44)$$

Our aim is to prove that given any time T, we can choose N sufficiently large so that

$$E(Iu)(t) \lesssim N^{2(1-s)}, \quad t \in [0, T].$$

Since δ depends on $\|Iu\|_{H^1}$, which is controlled by $\sqrt{E(Iu)} + \|g\|_{L^2}$, we can choose

$$\delta \approx N^{-4(1-s)-} \quad (4.45)$$

in Proposition 4.10 and Proposition 4.11 for each application below.

Using Proposition 4.11, we have on $[0, \delta]$

$$E(Iu) \leq E(Ig) + C\delta^{\frac{3}{4}-} N^{-1+} \|Iu\|_{X_\delta^{1,\frac{1}{2}+}}^6 \leq E(Ig) + C\delta^{\frac{3}{4}-} N^{-1+} N^{6(1-s)}.$$

To cover any time interval $[0, T]$, we need $\frac{T}{\delta}$ steps and thus

$$E(Iu) \leq E(Ig) + CN^{-1+} N^{6(1-s)} T \delta^{-\frac{1}{4}+} \lesssim N^{2(1-s)} + N^{-1+} N^{7(1-s)} T,$$

where we used (4.44) and (4.45). If we now pick N sufficiently large so that

$$TN^{5(1-s)-1+} \approx 1,$$

we obtain that

$$E(Iu) \lesssim N^{2(1-s)}.$$

Notice that this requires $s > \frac{4}{5}$, which completes the proof.

This argument also yields a growth bound for $\|u\|_{H^s}$ as follows. With the choice of N above, we have

$$E(Iu) \lesssim T^{\frac{2(1-s)}{5s-4-}}$$

and thus

$$\|u(T)\|_{H^s} \lesssim \left(\|g\|_{L^2}^2 + E(Iu)(T)\right)^{\frac{1}{2}} \lesssim C_{\|g\|_{H^s(\mathbb{T})}} T^{\frac{(1-s)}{5s-4-}}.$$

□

Remark 4.14 As we mentioned before, Theorem 4.13 is not optimal. In particular, one can improve the bound on the increment of the modified energy in Proposition 4.11, and thus weaken the regularity requirement in Theorem 4.13. Our intention in this chapter is to give a self-contained exposition of the I-method. To achieve this goal with minimal difficulty, we choose to present the main steps when the regularity index is very close to the energy level. This simplifies the argument considerably.

It remains to prove Proposition 4.11.

Proof of Proposition 4.11 Recall by (4.39) that

$$|E(Iu)(\delta) - E(Iu)(0)| \leq \left| \int_0^\delta \int_{\mathbb{T}} \overline{Iu_{xx}} \, (F(Iu) - IF(u)) \, dxdt \right| + \left| \int_0^\delta \int_{\mathbb{T}} \overline{IF(u)} \, (F(Iu) - IF(u)) \, dxdt \right|,$$

with $F(u) = |u|^4 u$. Therefore, to complete the proof it suffices to show that

$$\left| \int_0^\delta \int_{\mathbb{T}} \overline{Iu_{xx}} \left(|Iu|^4 Iu - I|u|^4 u \right) dxdt \right| \lesssim N^{-1+} \|Iu\|^6_{X_\delta^{1,\frac{1}{2}+}} \tag{4.46}$$

and

$$\left| \int_0^\delta \int_{\mathbb{T}} \overline{I(|u|^4 u)} \left(|Iu|^4 Iu - I(|u|^4 u) \right) dxdt \right| \lesssim \delta N^{-2+} \|Iu\|^{10}_{X_\delta^{1,\frac{1}{2}+}}. \tag{4.47}$$

We start with the estimate (4.47). First, we have the following remarks to fix some notation simplifying the proof.

Remark 4.15 (1) We write m_i for $m(k_i)$, m_{ij} for $m(k_i - k_j)$, and similarly

$$m_{j_1 j_2 \cdots j_n} = m\left(k_{j_1} - k_{j_2} + \cdots \pm k_{j_n}\right).$$

(2) We denote the Littlewood–Paley projections (in x variable) on the Fourier side as $\widehat{P_N u} = u_N$.

(3) In the proof, we pass to the Fourier side only in the space variable, and we assume that $\widehat{u}(t,k)$ is real and nonnegative. The proof for general u follows from this since we are working with L^2 based spaces; see Exercise 4.12.

We will first consider the contributions of fixed Littlewood–Paley projections (in x variable) of the 10 functions in (4.47), which can be written by Plancherel in the x variable using the non-negativity of $\widehat{u}$ as

$$\int_0^\delta \sum_{\sum_{j=0}^9 (-1)^j k_j = 0} m_{01234} \, (m_{56789} - m_5 \cdots m_9) \, u_{N_0}(t,k_0) u_{N_1}(t,k_1) \ldots u_{N_9}(t,k_9) dt. \tag{4.48}$$

Let $N_{\max} := \max_{j \in 0,\ldots 9} N_j$. We claim that

$$|(4.48)| \lesssim \delta N^{-2+} N_{\max}^{0-} \|Iu\|^2_{L_t^\infty H_x^1} \|u\|^8_{L_t^\infty L_x^\infty}. \tag{4.49}$$

Using Sobolev embedding, (4.37), and Lemma 3.9, we have

$$\|u\|_{L_t^\infty L_x^\infty} \lesssim \|u\|_{L_t^\infty H_x^s} \lesssim \|Iu\|_{L_t^\infty H_x^1} \lesssim \|Iu\|_{X^{s,\frac{1}{2}+}}.$$

Therefore, the claim implies that

$$|(4.47)| \lesssim \delta N^{-2+}\|Iu\|^{10}_{X^{s,\frac{1}{2}+}} \sum_{N_0,\ldots,N_9} N_{\max}^{0-} \lesssim \delta N^{-2+}\|Iu\|^{10}_{X^{s,\frac{1}{2}+}}.$$

It remains to prove (4.49). We remark that the I operator acts like an identity on $P_K u$ if $K \ll N$. Therefore, if $N_{\max}$ is much smaller than N, then (4.48) is identically zero. We can thus assume that $N_{\max} \gtrsim N$. We also denote the second largest frequency by N_{med}. In addition, we use the analogous notation $k_{\max}$ and k_{med}. Since $\sum_{j=0}^{9}(-1)^j k_j = 0$ and $|k_{\max}| \approx N_{\max} \gtrsim N$, we have

$$N_{\text{med}} \approx |k_{\text{med}}| \approx |k_{\max}| \gtrsim N.$$

Thus

$$N_{\max} \approx N_{\text{med}} \gtrsim N.$$

Using the bound $0 \le m \le 1$, we have

$$\left| m_{01234}\,(m_{56789} - m_5 \cdots m_9) \right| \lesssim 1.$$

Therefore, we can estimate the absolute value of the integrand in (4.48) by (suppressing the t dependence)

$$\sum_{\sum_{j=0}^{9}(-1)^j k_j = 0} \prod_{\ell=0}^{9} u_{N_\ell}(k_\ell).$$

The last pointwise estimate we need is the following

$$\frac{1}{m_{\max} N_{\max}} \approx \frac{N^{s-1}}{N_{\max}^{s-1} N_{\max}} \lesssim N^{-1+} N_{\max}^{0-},$$

where we used $N_{\max} \gtrsim N$. The same bound holds for m_{med}. Therefore

$$\begin{aligned} &\sum_{\sum_{j=0}^{9}(-1)^j k_j = 0} \prod_{\ell=0}^{9} u_{N_\ell}(k_\ell) \\ &\lesssim N^{-2+} N_{\max}^{0-} \sum_{\sum_{j=0}^{9}(-1)^j k_j = 0} m_{\max} N_{\max} u_{N_{\max}} m_{med} N_{med} u_{N_{med}} \prod_{\ell \neq \ell_{\max}, \ell_{\text{med}}} u_{N_\ell} \\ &= N^{-2+} N_{\max}^{0-} \sum_{\sum_{j=0}^{9}(-1)^j k_j = 0} \mathcal{F}(JP_{N_{\max}} Iu)\mathcal{F}(JP_{N_{\text{med}}} Iu) \prod_{\ell \neq \ell_{\max}, \ell_{\text{med}}} \mathcal{F}(P_{N_\ell} u). \end{aligned}$$

By the convolution structure and Plancherel, we can rewrite this (omitting the complex conjugates) as

$$\begin{aligned}N^{-2+}N_{\max}^{0-}\left|\int_{\mathbb{T}}(JP_{N_{\max}}Iu)(JP_{N_{\text{med}}}Iu)\prod_{\ell\neq\ell_{\max},\ell_{\text{med}}}P_{N_\ell}u\,dx\right| \\ \leq N^{-2+}N_{\max}^{0-}\|JP_{N_{\max}}Iu\|_{L_x^2}\|JP_{N_{\text{med}}}Iu\|_{L_x^2}\prod_{\ell\neq\ell_{\max},\ell_{\text{med}}}\|P_{N_\ell}u\|_{L_x^\infty} \\ \lesssim N^{-2+}N_{\max}^{0-}\|Iu\|_{H_x^1}^2\|u\|_{L_x^\infty}.\end{aligned}$$

In the last inequality, we used the bound

$$\sup_N\|P_Nu\|_{L^p}\lesssim\|u\|_{L^p},\quad 1\leq p\leq\infty.$$

This finishes the proof of (4.49) and hence the proof of (4.47).

We now prove the inequality (4.46). This is slightly harder since two derivatives appear on the left-hand side. Taking advantage of the estimate

$$\|(Iu)_{xx}\|_{X^{-1,\frac{1}{2}+}}\lesssim\|Iu\|_{X^{1,\frac{1}{2}+}},$$

it suffices to prove that

$$\left|\int_0^\delta\int_{\mathbb{T}}\overline{Iu}\left(|Iu|^4Iu-I(|u|^4u)\right)dxdt\right|\lesssim N^{-1+}\|Iu\|_{X_\delta^{-1,\frac{1}{2}+}}\|Iu\|_{X_\delta^{1,\frac{1}{2}+}}^5.$$

Again, we consider the Littlewood–Paley pieces. As before, it suffices to prove (after replacing Iu with u) that

$$\begin{aligned}\left|\int_0^\delta\sum_{\sum_{j=1}^6(-1)^jk_j=0}\left(\frac{m_2\cdots m_6-m_{23456}}{m_2\cdots m_6}\right)u_{N_1}(t)u_{N_2}(t)...u_{N_6}(t)dt\right| \\ \lesssim\delta^{\frac{3}{4}-}N_{\max}^{0-}N^{-1+}\|u\|_{X^{-1,1/2+}}\|u\|_{X^{1,1/2+}}^5\end{aligned}\qquad(4.50)$$

Notice that the left-hand side is symmetric with respect to $N_2\cdots N_6$, thus we can assume that

$$N_2\geq N_3\geq...\geq N_6.$$

As in the proof of (4.47), the largest two indices must be comparable since $\sum_{j=1}^6(-1)^jk_j=0$. Therefore, we always have $N_1\lesssim N_2$, and if $N_1\ll N_2$, then we must have $N_3\approx N_2$. We thus have the following cases:

Case 1: $N\gg N_2$,

Case 2: $N_1\approx N_2\gtrsim N\gg N_3$,

Case 3: $N_2\approx N_3\gtrsim N$ and $N_1\ll N_2$,

Case 4: $N_1\approx N_2\geq N_3>N$.

Case 1: $N \gg N_2$.
In this case, we have

$$\frac{m_{23456} - m_2 ... m_6}{m_2 ... m_6} = 0,$$

and hence there is nothing to prove.

Case 2: $N_1 \approx N_2 \gtrsim N \gg N_3$.
We have $m_3 = m_4 = m_5 = m_6 = 1$, and $m_1 = m_{23456}$. Therefore

$$\left| \frac{m_2 \cdots m_6 - m_{23456}}{m_2 \cdots m_6} \right| = \left| \frac{m_1 - m_2}{m_2} \right| \lesssim \frac{N_3}{N_2}.$$

In the last inequality, we used the mean value theorem together with the fact that

$$\left| \frac{\nabla m(\xi)}{m(\xi)} \right| \lesssim |\xi|^{-1}.$$

The bound on the multiplier can be considered as a gain of one derivative, and it is a by-product of the smoothness of the multiplier m. In the high–low decomposition method, since we used sharp frequency cut-offs, this gain is not available. Instead, we relied on a nonlinear smoothing bound, which gains less than half a derivative.

Using the multiplier bound above and $N_{\max} \approx N_1 \approx N_2 \gtrsim N$, we estimate

$$\begin{aligned}
&\left| \int_0^\delta \sum_{\sum_{j=1}^6 (-1)^j k_j = 0} \left(\frac{m_2 \cdots m_6 - m_{23456}}{m_2 \cdots m_6} \right) u_{N_1} u_{N_2} \ldots u_{N_6} dt \right| \\
&\quad \lesssim \int_0^\delta \sum_{\sum_{j=1}^6 (-1)^j k_j = 0} \frac{N_3}{N_2} u_{N_1} u_{N_2} \ldots u_{N_6} dt \\
&\quad \lesssim N_{\max}^{0-} N^{-1+} \int_0^\delta \sum_{\sum_{j=1}^6 (-1)^j k_j = 0} \frac{u_{N_1}}{N_1} N_2 u_{N_2} N_3 u_{N_3} u_{N_4} u_{N_5} u_{N_6} dt. \qquad (4.51)
\end{aligned}$$

Now by undoing Plancherel's theorem, and using Hölder's inequality, we bound this by

$$\begin{aligned}
&\delta^{\frac{1}{2}} N_{\max}^{0-} N^{-1+} \|J^{-1} u\|_{L_t^\infty L_x^2} \|Ju\|_{L_t^4 L_x^4}^2 \|u\|_{L_t^\infty L_x^\infty}^3 \\
&\quad \lesssim \delta^{\frac{1}{2}} N_{\max}^{0-} N^{-1+} \|u\|_{X^{-1,\frac{1}{2}+}} \|u\|_{X^{1,\frac{3}{8}}}^2 \|u\|_{X^{\frac{1}{2}+,\frac{1}{2}+}}^3 \\
&\quad \lesssim \delta^{\frac{3}{4}-} N_{\max}^{0-} N^{-1+} \|u\|_{X^{-1,\frac{1}{2}+}} \|u\|_{X^{1,\frac{1}{2}+}}^5 .
\end{aligned}$$

Here we used Sobolev embedding, Lemma 3.9, Lemma 3.11, and the improvement of the Strichartz estimate Theorem 3.25

$$\|u\|_{L_t^4 L_x^4} \lesssim \|u\|_{X^{0,\frac{3}{8}}} .$$

This completes the proof of this case.

Case 3: $N_2 \approx N_3 \gtrsim N$ and $N_1 \ll N_2$.

In this case, we use the crude estimate

$$\left|\frac{m_2\cdots m_6 - m_{23456}}{m_2\cdots m_6}\right| \lesssim 1 + \frac{m_1}{m_2\ldots m_6}. \tag{4.52}$$

We ignore the contribution of constant 1 term, which is easier to handle. Using

$$m_j N_j \gtrsim N_2^s N^{1-s}, \quad j = 2, 3,$$

and

$$m_1 N_1 \lesssim N_1^s \min\left(N_1^{1-s}, N^{1-s}\right),$$

we have

$$\frac{m_1}{m_2\ldots m_6} \lesssim \frac{N_2 N_3}{N_1} \frac{N_1^s \min\left(N_1^{1-s}, N^{1-s}\right)}{N_2^{2s} N^{2-s} m_4 m_5 m_6}. \tag{4.53}$$

We will consider two subcases: $N \gtrsim N_4 \geq N_5 \geq N_6$ and $N_4 \geq N_5 \geq N_6 \gtrsim N$. The other subcases can be handled similarly. If $N \gtrsim N_4 \geq N_5 \geq N_6$, we have $m_4 \approx m_5 \approx m_6 \approx 1$, and hence

$$|(4.53)| \lesssim \frac{N_2 N_3}{N_1} N_2^{0-} N^{-1+}.$$

Therefore, in this case

$$|(4.50)| \lesssim N_{\max}^{0-} N^{-1+} \int_0^\delta \sum_{\sum_{j=1}^6 (-1)^j k_j = 0} \frac{u_{N_1}}{N_1} N_2 u_{N_2} N_3 u_{N_3} u_{N_4} u_{N_5} u_{N_6} dt.$$

This is identical to the right-hand side of (4.51), and the proof can be completed as above.

If $N_4 \geq N_5 \geq N_6 \gtrsim N$, using the definition of m, we have (for $s > \frac{1}{2}$)

$$\frac{1}{m_j} \approx \frac{N^{s-1}}{N_j^{s-1}} \lesssim N^{-\frac{1}{2}+} N_j^{\frac{1}{2}-}, \quad j = 4, 5, 6.$$

Therefore

$$|(4.53)| \lesssim \frac{N_2 N_3 (N_4 N_5 N_6)^{\frac{1}{2}-}}{N_1} \frac{N_1^s \min\left(N_1^{1-s}, N^{1-s}\right) N^{-\frac{3}{2}+}}{N_2^{2s} N^{2-s}}$$

$$\lesssim \frac{N_2 N_3 (N_4 N_5 N_6)^{\frac{1}{2}-}}{N_1} N_{\max}^{0-} N^{-\frac{5}{2}+}.$$

Thus

$$|(4.50)| \lesssim N_{\max}^{0-} N^{-\frac{5}{2}+} \int_0^\delta \sum_{\sum_{j=1}^6 (-1)^j k_j = 0} \frac{u_{N_1}}{N_1} N_2 u_{N_2} N_3 u_{N_3} N_4^{\frac{1}{2}-} u_{N_4} N_5^{\frac{1}{2}-} u_{N_5} N_6^{\frac{1}{2}-} u_{N_6} dt.$$

This yields the required bound by proceeding as in the proof of (4.51) and using Sobolev embedding as follows

$$\|J^{\frac{1}{2}-} u\|_{L_t^\infty L_x^\infty} \lesssim \|u\|_{L_t^\infty H_x^1} \lesssim \|u\|_{X^{1,\frac{1}{2}+}}.$$

Case 4: $N_1 \approx N_2 \geq N_3 > N$.

In this case, we use the crude estimate (4.52), ignore the contribution of constant 1 term, and also restrict ourselves to the subcase $N \gtrsim N_4 \geq N_5 \geq N_6$.

In this case, instead of (4.50) we can only prove that

$$\left| \int_0^\delta \sum_{\sum_{j=1}^6 (-1)^j k_j = 0} \left(\frac{m_2 \cdots m_6 - m_{23456}}{m_2 \cdots m_6} \right) u_{N_1}(t) u_{N_2}(t) \ldots u_{N_6}(t) dt \right|$$
$$\lesssim \delta^{\frac{3}{4}-} N_3^{0-} N^{-1+} \|P_{N_1} u\|_{X^{-1,1/2+}} \|P_{N_2} u\|_{X^{1,1/2+}} \|u\|_{X^{1,1/2+}}^4. \quad (4.54)$$

Assuming (4.54), the sum over dyadic $N_3 \geq N_4 \geq N_5 \geq N_6$ can be bounded as before using the decay N_3^{0-}. Then the sum over $N_1 \approx N_2$ is bounded by

$$\delta^{\frac{3}{4}-} N^{-1+} \|u\|_{X^{1,1/2+}}^4 \sum_{N_1 \approx N_2} \|P_{N_1} u\|_{X^{-1,1/2+}} \|P_{N_2} u\|_{X^{1,1/2+}}$$
$$\lesssim \delta^{\frac{3}{4}-} N^{-1+} \|u\|_{X^{1,1/2+}}^4 \left[\sum_{N_1} \|P_{N_1} u\|_{X^{-1,1/2+}}^2 \right]^{1/2} \left[\sum_{N_2} \|P_{N_2} u\|_{X^{1,1/2+}}^2 \right]^{1/2}$$
$$\lesssim \delta^{\frac{3}{4}-} N^{-1+} \|u\|_{X^{-1,1/2+}} \|u\|_{X^{1,1/2+}}^5.$$

In the first inequality, we used fact that there are only finitely many dyadic $N_1 \approx N_2$ for fixed N_2.

To prove (4.54), first note that $m_1 \approx m_2$ and $m_4 = m_5 = m_6 = 1$. We have

$$\frac{m_1}{m_2 \cdots m_6} \approx \frac{1}{m_3} \lesssim \frac{N_3^{1-}}{N^{1-}},$$

where we used the fact that for any $0 < s < 1$ and $N_3 \gtrsim N$ we have that

$$m(N_3) N_3^{1-} \gtrsim N^{1-}.$$

Thus, the left-hand side of (4.54) is estimated (using the fact that $N_1 \approx N_2$) by

$$N_3^{0-} N^{-1+} \int_0^\delta \sum_{\sum_{i=1}^6 k_i = 0} N_1^{-1} u_{N_1} N_2 u_{N_2} N_3 u_{N_3} u_{N_4} u_{N_5} u_{N_6} dt.$$

Now by undoing Plancherel's theorem, and using Hölder's inequality as in the previous case, we bound this by

$$\begin{aligned}
&\delta^{\frac{1}{2}} N_3^{0-} N^{-1+} \|J^{-1} P_{N_1} u\|_{L_t^\infty L_x^2} \|J P_{N_2} u\|_{L_t^4 L_x^4} \|J u\|_{L_t^4 L_x^4} \|u\|_{L_t^\infty L_x^\infty}^3 \\
&\quad \lesssim \delta^{\frac{1}{2}} N_3^{0-} N^{-1+} \|P_{N_1} u\|_{X^{-1,\frac{1}{2}+}} \|P_{N_2} u\|_{X^{1,\frac{3}{8}}} \|u\|_{X^{1,\frac{3}{8}}} \|u\|_{X^{\frac{1}{2}+,\frac{1}{2}+}}^3 \\
&\qquad \lesssim \delta^{\frac{3}{4}-} N_3^{0-} N^{-1+} \|P_{N_1} u\|_{X^{-1,\frac{1}{2}+}} \|P_{N_2} u\|_{X^{1,\frac{1}{2}+}} \|u\|_{X^{1,\frac{1}{2}+}}^4 .
\end{aligned}$$

This finishes the proof of the proposition. □

In the focusing case

$$\begin{cases} iu_t + u_{xx} + |u|^4 u = 0, & t \in \mathbb{R}, \ x \in \mathbb{T}, \\ u(0,\cdot) = g(\cdot) \in H^s(\mathbb{T}) \end{cases} \tag{4.55}$$

the problem is similarly locally wellposed for any $s > 0$; see Bourgain [15, 20] and Theorem 3.31. We also have the conservation of mass

$$\|u(t)\|_{L^2(\mathbb{T})} = \|g\|_{L^2(\mathbb{T})},$$

and the conservation of energy

$$\begin{aligned}
E(u)(t) &= \frac{1}{2} \int_{\mathbb{T}} |u_x(t,x)|^2 \, dx - \frac{1}{6} \int_{\mathbb{T}} |u(t,x)|^6 \, dx \\
&= E(u)(0) = \frac{1}{2} \int_{\mathbb{T}} |g_x(x)|^2 \, dx - \frac{1}{6} \int_T |g(x)|^6 \, dx.
\end{aligned}$$

Note that by Gagliardo–Nirenberg inequality one can control $\|u\|_{H^1}$ by the energy and mass provided that $\|g\|_{L^2}$ is sufficiently small. Indeed

$$\begin{aligned}
E(u)(t) &= \frac{1}{2} \|u_x(t,x)\|_{L^2}^2 - \frac{1}{6} \|u(t,x)\|_{L^6}^6 \\
&\geq \|u_x(t,x)\|_{L^2}^2 \left(\frac{1}{2} - \frac{C}{6} \|u(t)\|_{L^2}^4 \right) \\
&= \|u_x(t,x)\|_{L^2}^2 \left(\frac{1}{2} - \frac{C}{6} \|g\|_{L^2}^4 \right) \gtrsim \|u_x(t,x)\|_{L^2}^2
\end{aligned}$$

provided that $\|g\|_{L^2}$ is small. Here C is the constant in Gagliardo–Nirenberg inequality. Therefore, the problem is globally wellposed in H^1 under this condition.

Furthermore, since

$$\|Iu\|_{L^2} \leq \|u\|_{L^2} = \|g\|_{L^2},$$

one has

$$\|Iu\|_{H^1}^2 \lesssim E(Iu) + \|g\|_{L^2}^2 .$$

Therefore, the assertion of Theorem 4.13 remains valid in the focusing case provided that $\|g\|_{L^2}$ is sufficiently small.

Exercises

4.1 (a) Prove that

$$\left|k^4 - m^4 + n^4 - (k - m + n)^4\right| \gtrsim |k - m||m - n|\left(k^2 + m^2 + n^2\right).$$

(b) Using part a, prove that a smoothing statement similar to Theorem 4.1 hold for the cubic fourth order NLS on the torus

$$\begin{cases} iu_t + \partial_x^4 u \pm |u|^2 u = 0, & x \in \mathbb{T}, \quad t \in \mathbb{R}, \\ u(0, \cdot) = g(\cdot) \in H^s(\mathbb{T}), \end{cases}$$

for $s \geq 0$ and $a < \min(2s + 1, \frac{3}{2})$.

4.2 Prove that the inequality (4.9) fails for any $s \in \mathbb{R}$ and for any $a > 0$ by taking

$$\widehat{u}(\tau, n) = \delta(n - M)\chi_{[-1,1]}(\tau - n^3), \quad M \text{ large, and}$$

$$\widehat{v}(\tau, n) = \delta(n - 1)\chi_{[-1,1]}(\tau - n^3).$$

4.3 (a) Show that if $g \in L^2(\mathbb{T})$ with $\widehat{g}(0) = 0$, then the function h_t with the Fourier sequence

$$\widehat{h_t}(k) = \sum_{k_1+k_2=k} \frac{\widehat{g}(k_1)\widehat{g}(k_2)}{kk_1}\left(e^{-3ik_1k_2kt} - 1\right)$$

is in $H^1(\mathbb{T})$.

(b) Note that the analogous argument for the first Picard iterate for the KdV equation on $\mathbb{R}$ does not imply any smoothing.

4.4 (a) Show that for any integer $k \geq 2$ the first Picard iterate of the equation

$$\begin{cases} u_t + \partial_x^{2k+1} u + uu_x = 0, & x \in \mathbb{R}, \quad t \in \mathbb{R}, \\ u(0, x) = g(x) \in L^2(\mathbb{R}) \end{cases}$$

has smoothing.

(b) Prove a nonlinear smoothing statement for this equation as in Theorem 4.1.

4.5 Consider the defocusing modified KdV equation (mKdV)

$$v_t + v_{xxx} = 6v^2 v_x \tag{4.56}$$

$$v(0, x) = g(x) \in H^s(\mathbb{T}), \quad s > 1/2.$$

(a) Show that the mKdV equation satisfies both mean and L^2 conservation.

(b) Define the Miura transform as

$$\mathbf{M}v := -6(v_x + v^2).$$

Prove that $M : H^s(\mathbb{T}) \to H^{s-1}(\mathbb{T})$ for $s > \frac{1}{2}$.

(c) Prove that if w solves (4.56), then $u = \mathbf{M}w$ solves the KdV equation

$$u_t + u_{xxx} + uu_x = 0$$

with the initial data $u(0, x) = \mathbf{M}g(x)$.

(d) This exercise together with Theorem 4.3 can be used to obtain a smoothing theorem for the mKdV equation, see [56].

4.6 Prove Theorem 4.9 for a larger range of s when $V = 0$.

4.7 In this exercise we describe how one obtains the wellposedness of the BBM equation

$$\begin{cases} u_t - u_{txx} + u_x + uu_x = 0, & x \in \mathbb{R},\ t \in \mathbb{R}, \\ u(0, x) = g(x) \in H^s(\mathbb{R}), & s \geq 0. \end{cases}$$

(a) Show that the smooth solutions satisfy the conservation law

$$E(u(t)) := \int_{\mathbb{R}} u^2 dx + \int_{\mathbb{R}} u_x^2 dx = E(g).$$

(b) Show that for any $s \geq 0$

$$\left\| \frac{\partial_x}{1 - \partial_{xx}} \left(u^2\right) \right\|_{H^s} \lesssim \|u\|_{H^s}^2.$$

(c) Obtain local wellposedness in $C_t^0 H_x^s$ for any $s \geq 0$ with the local existence time depending on $\|g\|_{H^s}$.

(d) Obtain global wellposedness in H^s, $s \geq 1$. In fact, global wellposedness holds in L^2 by a variation of the high–low decomposition method of Bourgain; see the paper [14] by Bona–Tzvetkov. This result is optimal.

4.8 This exercise discusses smoothing for the KdV equation on $\mathbb{R}$. In [62], Fonseca–Linares–Ponce proved a smoothing statement for the focusing modified KdV equation. Their proof is valid also for the defocusing case. Using their statement and the Miura transform, prove that if the initial data

$$g = \phi^2 + \phi_x$$

for some ϕ, then the KdV equation on $\mathbb{R}$ with data g has smoothing.

4.9 Prove that the I operator (4.35) satisfies:
(a) for $0 \leq s \leq 1$, we have

$$\|u\|_{H^s} \lesssim \|Iu\|_{H_1} \lesssim N^{1-s}\|u\|_{H^s}.$$

(b) Prove that

$$\|Iu\|_{L^p(\mathbb{R}^n)} \lesssim \|u\|_{L^p(\mathbb{R}^n)},$$

for $1 < p < \infty$.

4.10 Let m be the multiplier defined in (4.34). Prove for any $0 < s < 1$ that on the hyperplane $\sum_{j=1}^{6} k_j = 0$ we have

$$\frac{m(k_6)\langle k_6\rangle^{1-s}}{\prod_{j=1}^{5} m(k_j)\langle k_j\rangle^{1-s}} \lesssim 1.$$

4.11 Prove the inequality (4.42) using the Cauchy–Schwarz inequality.

4.12 Prove Proposition 4.11 without assuming that the Fourier sequence of u is real and nonnegative.

5

Applications of smoothing estimates

5.1 Bounds for higher order Sobolev norms

In this section, we discuss growth bounds on the Sobolev norms of the solutions. Recall that for the cubic NLS equation, the complete integrability implies that all Sobolev norms H^k, $k \in \mathbb{N}$, are bounded in time; see e.g. Zakharov–Shabat [157], Wadati–Sanuki–Konno [153], and Yang [155]. Also recall that, by the recent results of Kappeler–Schaad–Topalov [80], the H^s norm remains bounded for any real $s \geq 1$ in the defocusing case. In [17, 20], Bourgain introduced a method for obtaining a priori bounds for the higher order Sobolev norms for equations lacking suitable conservation laws. In [130], Staffilani considered the cubic NLS equation with a time independent potential in front of the nonlinearity, and obtained polynomial growth bounds in this nonintegrable case. Here, we present the method for the fractional NLS equation on the torus

$$\begin{cases} iu_t + D^{2\alpha}u \pm |u|^2u = 0, & x \in \mathbb{T}, \quad t \in \mathbb{R}, \\ u(0,x) = g(x) \in H^s(\mathbb{T}), \end{cases} \tag{5.1}$$

where $\alpha \in (\frac{1}{2}, 1]$. We refer the reader to Demirbaş–Erdoğan–Tzirakis [51] for the local wellposedness theory in $H^s(\mathbb{T})$, $s > \frac{1-\alpha}{2}$. For the wellposedness theory of fractional Schrödinger on $\mathbb{R}$, see Cho–Hwang–Kwon–Lee [35]. The local theory follows from a trilinear estimate in $X^{s,b}$ spaces for $s > \frac{1-\alpha}{2}$ and $b > \frac{1}{2}$. The $X^{s,b}$ norm for the fractional NLS is defined as

$$\|u\|_{X^{s,b}} = \left\| e^{-itD^{2\alpha}}u \right\|_{H^s_x H^b_t} = \left\| \langle k\rangle^s \left\langle \tau - k^{2\alpha}\right\rangle^b \widehat{u}(\tau,k) \right\|_{L^2_\tau \ell^2_k}.$$

Global wellposedness in $H^\alpha(\mathbb{T})$ level follows immediately, since the L^2 norm conservation and the energy conservation

$$E(u)(t) = \int_{\mathbb{T}} |D^\alpha u(t,x)|^2 \pm \frac{1}{2}\int_{\mathbb{T}} |u(t,x)|^4 = E(u)(0),$$

provide a uniform in time bound for the $H^\alpha(\mathbb{T})$ norm. The uniform bound holds also in the focusing case, since the equation is mass and energy sub-critical. This follows from the Gagliardo–Nirenberg inequality (Lemma 1.7)

$$\|u\|_{L^4}^4 \lesssim \|D^\alpha u\|_{L^2}^{\frac{1}{\alpha}} \|u\|_{L^2}^{4-\frac{1}{\alpha}},$$

which controls the potential energy via the kinetic energy $\|D^\alpha u\|_{L^2}$. One can then control the Sobolev norm of the solution for all times even in the focusing case since $\frac{1}{\alpha} < 2$; see Exercise 5.1.

For higher order Sobolev norms, we have the following bounds:

Theorem 5.1 *Consider the initial value problem* (5.1) *where* $\alpha \in (1/2, 1]$. *For any real* $s > \alpha$, *we have the following growth bound*

$$\|u(t)\|_{H^s(\mathbb{T})} \lesssim \begin{cases} \langle t \rangle, & s < 2\alpha - \frac{1}{2} \\ \langle t \rangle^{\frac{2s-2\alpha}{2\alpha-1}}, & s \geq 2\alpha - \frac{1}{2}, \end{cases}$$

where the implicit constants depend on the $H^s(\mathbb{T})$ *norm of the initial data.*

The proof of Theorem 5.1 will rely on a smoothing theorem for the nonresonant terms of the nonlinearity; see Proposition 5.2 below. Recall the identity (4.4) that follows from Plancherel's theorem and the conservation of L^2 norm

$$\widehat{|u|^2 u}(k) = \frac{1}{\pi}\|u\|_2^2 \widehat{u}(k) - |\widehat{u}(k)|^2 \widehat{u}(k) + \sum_{k_1 \neq k, k_2 \neq k_1} \widehat{u}(k_1)\overline{\widehat{u}(k_2)}\widehat{u}(k - k_1 + k_2)$$

$$=: P\widehat{u}(k) + \widehat{\rho(u)}(k) + \widehat{R(u)}(k), \quad (5.2)$$

where $P = \frac{1}{\pi}\|g\|_2^2$. We start with the following smoothing result for the second and third summands above. Note that this is a stronger smoothing statement then the estimates we obtained for the NLS and KdV equations since the right hand depends only linearly on the H^s norm, although it requires smoother initial data.

Proposition 5.2 *For* $\alpha \in (\frac{1}{2}, 1]$, $s \geq \alpha$, *and* $c \leq \alpha - \frac{1}{2}$, *we have*

$$\|\rho(u) + R(u)\|_{X^{s+c,-\frac{1}{2}}} \lesssim \|u\|_{X^{s,b}} \|u\|_{X^{\alpha,b}}^2,$$

for any $b > \frac{1}{2}$.

Proof We start with $R(u)$. First note that

$$\|R(u)\|_{X^{s+c,-\frac{1}{2}}} = \left\| \int_{\tau_1-\tau_2+\tau_3=\tau} \sum_{\substack{k_1-k_2+k_3=n \\ k_1\neq n,k_2}} \frac{\widehat{u}(\tau_1,k_1)\overline{\widehat{u}(\tau_2,k_2)}\widehat{u}(\tau_3,k_3)\langle n\rangle^{s+c}}{\langle \tau - n^{2\alpha}\rangle^{\frac{1}{2}}} \right\|_{L^2_\tau l^2_n}.$$

By denoting

$$v(\tau,n) = |\widehat{u}(\tau,n)|\langle n\rangle^{\alpha}\left\langle \tau - n^{2\alpha}\right\rangle^{b},$$

we get

$$\|R(u)\|_{X^{s+c,-\frac{1}{2}}} \leq \Bigg\| \int_{\tau_1-\tau_2+\tau_3-\tau=0} \sum_{\substack{k_1-k_2+k_3-n=0 \\ k_1\neq n,k_2}} \frac{\langle n\rangle^{s+c} v(\tau_1,k_1)v(\tau_2,k_2)v(\tau_3,k_3)}{\langle k_1\rangle^{\alpha}\langle k_2\rangle^{\alpha}\langle k_3\rangle^{\alpha}\left\langle \tau - n^{2\alpha}\right\rangle^{\frac{1}{2}}} \times \frac{1}{\left\langle \tau_1 - k_1^{2\alpha}\right\rangle^{b}\left\langle \tau_2 - k_2^{2\alpha}\right\rangle^{b}\left\langle \tau_3 - k_3^{2\alpha}\right\rangle^{b}} \Bigg\|_{L^2_\tau l^2_n}.$$

Using that $\langle n\rangle \lesssim \langle k_1\rangle + \langle k_2\rangle + \langle k_3\rangle$, we have

$$\|R(u)\|_{X^{s+c,-\frac{1}{2}}} \lesssim \Bigg\| \int_{\tau_1-\tau_2+\tau_3-\tau=0} \sum_{\substack{k_1-k_2+k_3-n=0 \\ k_1\neq n,k_2}} \frac{\langle n\rangle^{\alpha+c} v(\tau_1,k_1)v(\tau_2,k_2)v(\tau_3,k_3)}{\langle k_1\rangle^{\alpha}\langle k_2\rangle^{\alpha}\langle k_3\rangle^{\alpha}\left\langle \tau - n^{2\alpha}\right\rangle^{\frac{1}{2}}} \times \frac{(\langle k_1\rangle + \langle k_2\rangle + \langle k_3\rangle)^{s-\alpha}}{\left\langle \tau_1 - k_1^{2\alpha}\right\rangle^{b}\left\langle \tau_2 - k_2^{2\alpha}\right\rangle^{b}\left\langle \tau_3 - k_3^{2\alpha}\right\rangle^{b}} \Bigg\|_{L^2_\tau l^2_n}.$$

Thus, by the Cauchy–Schwarz inequality, we have

$$\begin{aligned}\|R(u)\|^2_{X^{s+c,-b'}} \lesssim & \int_{\mathbb{R}^3}\sum_{\mathbb{Z}^3}(\langle k_1\rangle + \langle k_2\rangle + \langle k_3\rangle)^{2s-2\alpha} v(\tau_1,k_1)^2 v(\tau_2,k_2)^2 v(\tau_3,k_3)^2 \\ & \times \sup_{\tau,n}\Bigg[\int_{\tau_1-\tau_2+\tau_3=\tau} \sum_{\substack{k_1-k_2+k_3=n \\ k_1\neq n,k_2}} \frac{\langle n\rangle^{2\alpha+2c}}{\langle k_1\rangle^{2\alpha}\langle k_2\rangle^{2\alpha}\langle k_3\rangle^{2\alpha}\left\langle \tau - n^{2\alpha}\right\rangle} \\ & \times \frac{1}{\left\langle \tau_1 - k_1^{2\alpha}\right\rangle^{2b}\left\langle \tau_2 - k_2^{2\alpha}\right\rangle^{2b}\left\langle \tau_3 - k_3^{2\alpha}\right\rangle^{2b}}\Bigg].\end{aligned}$$

The first line can be bounded by $\|u\|^2_{X^{s,b}}\|u\|^4_{X^{\alpha,b}}$. Integrating in τ variables using Exercise 3.12, we have

$$\|R(u)\|^2_{X^{s+c,-\frac{1}{2}}} \lesssim \|u\|^2_{X^{s,b}}\|u\|^4_{X^{\alpha,b}} \sup_n \sum_{\substack{k_1-k_2+k_3=n \\ k_1\neq n,k_2}} \frac{\langle n\rangle^{2\alpha+2c}}{\langle k_1\rangle^{2\alpha}\langle k_2\rangle^{2\alpha}\langle k_3\rangle^{2\alpha}\left\langle k_1^{2\alpha} - k_2^{2\alpha} + k_3^{2\alpha} - n^{2\alpha}\right\rangle}.$$

Hence, we need to show that

$$M_n = \sum_{\substack{k_1-k_2+k_3=n \\ k_1\neq n,k_2}} \frac{\langle n\rangle^{2\alpha+2c}}{\langle k_1\rangle^{2\alpha}\langle k_2\rangle^{2\alpha}\langle k_3\rangle^{2\alpha}\left\langle k_1^{2\alpha}-k_2^{2\alpha}+k_3^{2\alpha}-n^{2\alpha}\right\rangle}$$

is bounded in n.

Renaming the variables as

$$k_1 = n+j,\;\; k_2 = n+k+j,\;\; k_3 = n+k,$$

and using Lemma 2.13, it suffices to bound

$$\sum_{kj\neq 0} \frac{\langle n\rangle^{2\alpha+2c}}{\langle n+j\rangle^{2\alpha}\langle n+k\rangle^{2\alpha}\langle n+k+j\rangle^{2\alpha}\left\langle \frac{|kj|}{(|n|+|k|+|j|)^{2-2\alpha}}\right\rangle}.$$

For the terms with $0<|kj|\lesssim |n|^{2-2\alpha}$, since $|k|,|j|\ll |n|$, we have the bound

$$\lesssim \sum_{0<|kj|\lesssim |n|^{2-2\alpha}} \langle n\rangle^{-4\alpha+2c} \lesssim \langle n\rangle^{2c+2-6\alpha}\log(n),$$

which is bounded provided that $c<3\alpha-1$.

For the remaining terms, by symmetry in j and k, it is enough to consider the case $|k|\geq |j|$. We will consider three regions:

Region 1. $|k|\gg |n|$.
We bound the sum in this region by

$$\sum_{\substack{|k|\geq |j|>0 \\ |k|\gg |n|}} \frac{\langle n\rangle^{2\alpha+2c}|k|^{1-4\alpha}}{\langle n+j\rangle^{2\alpha}\langle n+k+j\rangle^{2\alpha}|j|}$$

$$\lesssim \sum_{\substack{j \\ |k|\gg |n|}} \frac{\langle n\rangle^{2c+1-2\alpha}}{\langle n+j\rangle^{2\alpha}\langle n+k+j\rangle^{2\alpha}|j|}$$

$$\lesssim \sum_{j} \frac{\langle n\rangle^{2c+1-2\alpha}}{\langle n+j\rangle^{2\alpha}\langle j\rangle} \lesssim \langle n\rangle^{2c-2\alpha},$$

which is bounded for $c\leq\alpha$. In the third inequality, we used Exercise 3.12.

Region 2. $|k|\approx |n|$.
In this region, we have the bound,

$$\sum_{\substack{|k|\geq|j|>0\\|k|\approx|n|}} \frac{\langle n\rangle^{2c+1}}{\langle n+k\rangle^{2\alpha}\langle n+j\rangle^{2\alpha}\langle n+k+j\rangle^{2\alpha}|j|}$$

$$\lesssim \sum_{|k|\approx|n|} \frac{\langle n\rangle^{2c+1-2\alpha}}{\langle n+k\rangle^{2\alpha}} \lesssim \langle n\rangle^{2c+1-2\alpha} \lesssim 1$$

for $c \leq \alpha - \frac{1}{2}$.

Region 3. $|k| \ll |n|$.

We have

$$\lesssim \sum_{\substack{|k|\geq|j|>0\\|k|\ll|n|}} \frac{\langle n\rangle^{2c+2-6\alpha}}{|kj|} \lesssim \langle n\rangle^{2c+2-6\alpha+},$$

which is bounded for $c < 3\alpha - 1$.

Finally, note that for $s \geq \alpha$ and $c < 2\alpha$, we have (see Exercise 5.2)

$$\|\rho(u)\|_{X^{s+c,-\frac{1}{2}}} \lesssim \|u\|_{X^{s,\frac{1}{2}+}} \|u\|^2_{X^{\alpha,\frac{1}{2}+}}. \tag{5.3}$$

□

We can now prove Theorem 5.1. By differentiating $\|u\|^2_{H^s}$ and using equation (5.1), we have

$$\partial_t \|u(t)\|^2_{H^s} = 2\Re \int_{\mathbb{T}} \overline{D^s u} D^s u_t \, dx = \mp 2\Im \int_{\mathbb{T}} \overline{D^s u}\, D^s(|u|^2 u)\, dx$$

Therefore

$$\|u(t)\|^2_{H^s} = \|g\|^2_{H^s} \mp 2\Im \int_0^t \int_{\mathbb{T}} \overline{D^s u}\, D^s(|u|^2 u) dx dt'.$$

By (5.2), we have

$$|u|^2 u = \frac{1}{\pi}\|g\|_2^2 u + \rho(u) + R(u).$$

Noting that the contribution of the first summand is zero, we have

$$\|u(t)\|^2_{H^s} = \|g\|^2_{H^s} \mp 2\Im \int_0^t \int_{\mathbb{T}} \overline{D^s u}\, D^s(\rho(u) + R(u)) dx dt'.$$

By Plancherel and Cauchy–Schwarz inequality in the space-time integral, for any $t \leq \delta$ we have the bound

$$\|u(t)\|^2_{H^s} - \|g\|^2_{H^s} \lesssim \|D^s u\|_{X_\delta^{0,\frac{1}{2}+}} \|D^s(\rho(u) + R(u))\|_{X_\delta^{0,-\frac{1}{2}-}}$$

$$\lesssim \|u\|_{X_\delta^{s,\frac{1}{2}+}} \|\rho(u) + R(u)\|_{X_\delta^{s,-\frac{1}{2}}}.$$

By Proposition 5.2 and the local wellposedness theory (for suitable δ), we have (for $c = \alpha - \frac{1}{2}$ and $s \geq 2\alpha - \frac{1}{2}$)

$$\|u(t)\|_{H^s}^2 - \|g\|_{H^s}^2 \lesssim \|u\|_{X_\delta^{s,\frac{1}{2}+}} \|u\|_{X_\delta^{s-c,\frac{1}{2}+}} \|u\|^2_{X_\delta^{\alpha,\frac{1}{2}+}}$$
$$\lesssim \|g\|_{H^s} \|g\|_{H^{s-c}} \|g\|^2_{H^\alpha} \lesssim \|g\|_{H^s}^{2-\frac{c}{s-\alpha}} \|g\|_{H^\alpha}^{2+\frac{c}{s-\alpha}}.$$

The last inequality follows by bounding the H^{s-c} norm by interpolating between H^α and H^s norms. Since the H^α norm remains bounded in time, for any $k \in \mathbb{N}$ we have

$$\|u((k+1)\delta)\|_{H^s} - \|u(k\delta)\|_{H^s} \lesssim \|u(k\delta)\|_{H^s}^{1-\frac{2\alpha-1}{2s-2\alpha}}. \tag{5.4}$$

Lemma 5.3 *Suppose a_k is a sequence of nonnegative numbers satisfying for some $\eta > 0$ and for all k*

$$a_{k+1} - a_k \leq C_1 a_k^{1-\eta}.$$

Then there exists C_2 such that for each k

$$a_k \leq C_2 k^{\frac{1}{\eta}}.$$

Proof By induction, it suffices to prove that

$$C_2 k^{\frac{1}{\eta}} + C_1 C_2^{1-\eta} k^{\frac{1-\eta}{\eta}} \leq C_2 (k+1)^{\frac{1}{\eta}}.$$

This follows from

$$1 + \frac{C_1}{k C_2^\eta} \leq \left(1 + \frac{1}{k}\right)^{\frac{1}{\eta}},$$

which in turn follows from the binomial expansion provided that C_2 is large. □

Using Lemma 5.3 and (5.4), for any s real with $s \geq 2\alpha - \frac{1}{2}$ we have

$$\|u(t)\|_{H^s} \lesssim \langle t \rangle^{\frac{2s-2\alpha}{2\alpha-1}}.$$

Notice that for $s < 2\alpha - \frac{1}{2}$, we can use

$$\|R(u)\|_{X^{s,-\frac{1}{2}}} \lesssim \|u\|^3_{X^{\alpha,\frac{1}{2}+}}.$$

This corresponds to the case $\eta = 1$ and yields the linear bound in time, finishing the proof of Theorem 5.1.

We note that one can also get growth bounds for higher order Sobolev norms by iterating the smoothing results we have in Section 4.1. We present this method in Section 5.4.2 below for the forced and damped KdV equation. Another method for obtaining such growth bounds is an upside-down version of the I-method; see Sohinger [128, 129] and Colliander–Kwon–Oh [45].

5.2 Almost everywhere convergence to initial data

In this chapter we consider the problem of almost everywhere convergence to initial data for linear and nonlinear equations on the torus. This problem was introduced by Carleson [26] in the case of the linear Schrödinger evolution on the real line. He proved that for any $g \in H^{\frac{1}{4}}(\mathbb{R})$, $e^{it\partial_{xx}}g$ converges to g almost everywhere on $\mathbb{R}$ as $t \to 0$. This result is optimal, see the paper of Dahlberg–Kenig [48]. To obtain this result, Carleson proved a maximal function estimate for the evolution. The Strichartz estimates we have in Chapter 2 can be used to obtain such maximal function estimates, as was observed by Moyua–Vega in [115] for the linear Schrödinger evolution on the torus. We present the method in the case of the linear fractional Schrödinger evolution.

Theorem 5.4 *Fix* $\alpha \in (\frac{1}{2}, 1]$. *For any* $s > \frac{1+\alpha}{4}$, *we have*

$$\left\| \sup_{t\in[0,1]} \left| e^{-itD^{2\alpha}} g \right| \right\|_{L^4_{x\in\mathbb{T}}} \lesssim \|g\|_{H^s}.$$

Note that the case $\alpha = 1$ follows immediately from the Sobolev embedding theorem.

Proof First assume that $\widehat{g} = 0$ outside $[-N, N]$. By differentiating $|e^{-itD^{2\alpha}} g|^4$ in time, we obtain

$$\left| e^{-itD^{2\alpha}} g \right|^4 = |g(x)|^4 + 4 \int_0^t \left| e^{-isD^{2\alpha}} g \right|^2 \Re \left(\overline{e^{-isD^{2\alpha}} g}\, \partial_s e^{-isD^{2\alpha}} g \right) ds.$$

Therefore

$$\left\| \sup_{t\in[0,1]} \left| e^{-itD^{2\alpha}} g \right| \right\|^4_{L^4_{x\in\mathbb{T}}} \lesssim \|g\|^4_{L^4} + \left\| e^{-itD^{2\alpha}} g \right\|^3_{L^4_{x,t\in\mathbb{T}}} \left\| \partial_t\, e^{-itD^{2\alpha}} g \right\|_{L^4_{x,t\in\mathbb{T}}}.$$

By Sobolev embedding, we have

$$\|g\|_{L^4} \lesssim N^{1/4} \|g\|_{L^2}.$$

By Theorem 2.12, we have

$$\left\| e^{-itD^{2\alpha}} g \right\|_{L^4_{x,t\in\mathbb{T}}} \lesssim N^{\frac{1-\alpha}{4}+} \|g\|_{L^2}.$$

Finally, since

$$\partial_t\, e^{-itD^{2\alpha}} g = -ie^{-itD^{2\alpha}} D^{2\alpha} g,$$

Theorem 2.12 also implies that

$$\left\| \partial_t\, e^{-itD^{2\alpha}} g \right\|_{L^4_{x,t\in\mathbb{T}}} \lesssim N^{\frac{1+7\alpha}{4}+} \|g\|_{L^2}.$$

Combining these estimates, we see that

$$\Big\| \sup_{t\in[0,1]} |e^{-itD^{2\alpha}} g| \Big\|_{L^4_{x\in\mathbb{T}}} \lesssim \big(N^{\frac14} + N^{\frac{1+\alpha}{4}+}\big)\|g\|_{L^2} \lesssim N^{\frac{1+\alpha}{4}+}\|g\|_{L^2}.$$

This yields the assertion of the theorem as in the proof of Theorem 2.9. □

Similarly, Theorem 2.8 and Theorem 2.10 yield

Theorem 5.5 *We have*

$$\Big\| \sup_{t\in[0,1]} \big|e^{it\partial_{xx}} g\big| \Big\|_{L^6_{x\in\mathbb{T}}} \lesssim \|g\|_{H^s}, \quad \textit{for any } s > \frac{1}{3}, \textit{ and}$$

$$\Big\| \sup_{t\in[0,1]} \big|e^{-t\partial_x^3} g\big| \Big\|_{L^{14}_{x\in\mathbb{T}}} \lesssim \|g\|_{H^s}, \quad \textit{for any } s > \frac{3}{7}.$$

Almost everywhere convergence to initial data follows from the maximal function estimates by a well-known approximation argument:

Corollary 5.6 *Fix $\alpha \in (\frac12, 1)$ and $s > \frac{1+\alpha}{4}$. For any $g \in H^s$, $e^{-itD^{2\alpha}} g$ converges to g almost everywhere as $t \to 0$. The analogous statements hold for the linear Schrödinger and Airy evolutions for $s > \frac13$ and $s > \frac37$ respectively.*

Proof We present the proof only for the fractional Schrödinger evolution. Note that by Sobolev embedding the statement holds for any H^1 data. Given $g \in H^s$, and $\epsilon > 0$, take $h \in H^1$ such that $\|g - h\|_{H^s} < \epsilon$. Write $g = f + h$. Note that for any $\lambda > 0$

$$\begin{aligned}
&\Big\{x : \limsup_{t\to 0} \big|e^{-itD^{2\alpha}} g(x) - g(x)\big| > \lambda\Big\} \\
&\qquad = \Big\{x : \limsup_{t\to 0} \big|e^{-itD^{2\alpha}} f(x) - f(x)\big| > \lambda\Big\} \\
&\qquad \subset \Big\{x : \sup_{t\in(0,1)} \big|e^{-itD^{2\alpha}} f(x)\big| > \frac{\lambda}{2}\Big\} \cup \Big\{x : |f(x)| > \frac{\lambda}{2}\Big\}.
\end{aligned}$$

Therefore, by Tchebychev's inequality and the maximal function estimate, we have

$$\begin{aligned}
&m\Big\{x : \limsup_{t\to 0} \big|e^{-itD^{2\alpha}} g(x) - g(x)\big| > \lambda\Big\} \\
&\qquad \le \Big(\frac{2}{\lambda} \Big\| \sup_{t\in(0,1)} \big|e^{-itD^{2\alpha}} f(x)\Big\|_{L^4}\Big)^4 + \Big(\frac{2}{\lambda}\|f(x)\|_{L^2}\Big)^2 \\
&\qquad \lesssim \Big(\frac{2}{\lambda}\epsilon\Big)^4 + \Big(\frac{2}{\lambda}\epsilon\Big)^2.
\end{aligned}$$

Since this holds for any $\epsilon > 0$, and $\lambda > 0$, the statement follows. □

Utilizing the smoothing estimates in Theorem 4.1 and Theorem 4.3, we can extend this corollary to the corresponding nonlinear equation. In the case of the nonlinear fractional NLS equation (5.1), we refer the reader to [51] for the following smoothing estimate:

Theorem 5.7 *Fix $\alpha \in (\frac{1}{2}, 1)$, $s > \frac{1-\alpha}{2}$, and $c < \min(2s + \alpha - 1, \alpha - \frac{1}{2})$. Let u be the solution of* (5.1) *with $g \in H^s(\mathbb{T})$. We have*

$$u - e^{itD^{2\alpha} - iPt} g \in C_t^0 H_x^{s+c}([0, \delta] \times \mathbb{T}),$$

where $[0, \delta]$ is the local existence interval and $P = \frac{1}{\pi}\|g\|_2^2$.

Note that this theorem follows from Proposition 5.2 in the case $s \geq \alpha$.

Corollary 5.8 *Fix $\alpha \in (\frac{1}{2}, 1)$ and $s > \max\left(\frac{1+\alpha}{4}, 1-\alpha\right)$. For any $g \in H^s(\mathbb{T})$, the solution of the nonlinear fractional NLS equation* (5.1) *converges to g almost everywhere as $t \to 0$.*

The analogous statements hold for the nonlinear Schrödinger (4.1) *and KdV evolutions* (4.10) *for $s > \frac{1}{3}$ and $s > \frac{3}{7}$ respectively.*

Proof We only discuss the case of the nonlinear fractional NLS equation. By Theorem 5.1 and Sobolev embedding, for $s > \max\left(\frac{1+\alpha}{4}, 1 - \alpha\right)$ and for $g \in H^s(\mathbb{T})$, we have

$$u - e^{itD^{2\alpha} - iPt} g \in C_t^0 H_x^{\frac{1}{2}+}([0, \delta] \times \mathbb{T}) \subset C_{t,x}^0.$$

Therefore

$$u(t, x) - e^{itD^{2\alpha} - iPt} g(x) \to 0$$

as $t \to 0$ for each $x \in \mathbb{T}$. This together with Corollary 5.6 yield the statement. □

5.3 Nonlinear Talbot effect

In this section, we extend some of the results of Section 2.3 on the Talbot effect for linear dispersive PDEs on $\mathbb{T}$ to the NLS and KdV equations on $\mathbb{T}$ using the smoothing estimates Theorem 4.1 and Theorem 4.3.

We should note that Olver [120] and Chen–Olver [33, 34] provided numerical simulations of the Talbot effect for a large class of dispersive equations, both linear and nonlinear. This behavior persists for both integrable and nonintegrable systems. In the case of linear equations with polynomial dispersion, they numerically confirmed the rational/irrational dichotomy that we discussed in Theorem 2.14 and Theorem 2.16 above. An important question that was raised in [33, 34] is the appearance of such phenomena in the case of nonpolynomial dispersion relations. The numerics demonstrate that the large wave number asymptotics of the dispersion relation plays the dominant role governing the qualitative features of the solutions. We should also note that in Zhang–Wen–Zhu–Xiao [158] the Talbot effect was observed experimentally in a nonlinear setting.

The smoothing theorems presented in Section 4.1 allow one [56, 59, 36] to extend some of the linear results we discussed above to the NLS and KdV evolutions. For example, combining Theorem 2.16 with the smoothing Theorem 4.1 (see Exercise 5.4), we have the following:

Corollary 5.9 *Let $g : \mathbb{T} \to \mathbb{R}$ be of bounded variation. Then for almost every t, the solution of the NLS equation on the torus*

$$\begin{cases} iu_t + u_{xx} + |u|^2 u = 0, & x \in \mathbb{T}, \quad t \in \mathbb{R}, \\ u(0, \cdot) = g(\cdot), \end{cases} \tag{5.5}$$

is in $C^\alpha(\mathbb{T})$ for $\alpha < \frac{1}{2}$, and both the real part and the imaginary part of the graph of u have fractal dimension $D \leq \frac{3}{2}$.

Moreover, if in addition $g \notin \bigcup_{\epsilon>0} H^{r_0+\epsilon}$ for some $r_0 \in [\frac{1}{2}, \frac{5}{8})$, then for almost all t both the real part and the imaginary part of the graph of u have fractal dimension $D \geq \frac{5}{2} - 2r_0$.

Note that for step function initial data ($r_0 = \frac{1}{2}$), we get $D = \frac{3}{2}$, justifying Berry's conjecture that we discussed in Section 2.3 for the cubic NLS equation.

Similarly, combining Theorem 2.16 with the smoothing Theorem 4.3, we have:

Corollary 5.10 *Let $g : \mathbb{T} \to \mathbb{R}$ be of bounded variation. Then for almost every t, the solution of the KdV equation on the torus*

$$\begin{cases} u_t + u_{xxx} + uu_x = 0, & x \in \mathbb{T}, \quad t \in \mathbb{R}, \\ u(0, \cdot) = g(\cdot) \end{cases}$$

is in $C^\alpha(\mathbb{T})$ for each $\alpha < \frac{1}{4}$. In particular, the dimension of the graph of the solution is $\leq \frac{7}{4}$ for almost all t.

Moreover, if in addition $g \notin \bigcup_{\epsilon>0} H^{r_0+\epsilon}$ for some $r_0 \in [\frac{1}{2}, \frac{5}{8})$, then for almost all t the graph of the solution have fractal dimension $D \geq \frac{9}{4} - 2r_0$.

We note that the simulations in Olver [120] and Chen–Olver [33, 34] were performed in the case when g is a step function, and that the corollaries above apply to that particular case.

Remark 5.11 We note that Oskolkov's result [121] and Theorem 4.1 imply that for bounded variation data g the solution $u(t,x)$ of (5.5) is a continuous function of x if $\frac{t}{2\pi}$ is an irrational number. Moreover, if g is also continuous, then $u \in C_t^0 C_x^0$. The same statement holds for the KdV evolution by Theorem 4.3.

5.4 Global attractors for dissipative and dispersive PDEs

Another application of the smoothing estimates we have in Section 4.1 is the existence and uniqueness of global attractors for dissipative and dispersive PDEs [57, 58]. In general, dissipative systems have bounded absorbing sets into which all solutions enter eventually. Notice that this is in contrast with conservative Hamiltonian systems where the orbits may fill the whole space or regions of it; see, e.g., Kuksin [99].

The problem of global attractors for nonlinear PDEs has generated considerable interest among engineers, physicists, and mathematicians in the last several decades. The theory is concerned with the description of the nonlinear dynamics as $t \to \infty$. The aim is to describe the long time asymptotics of the solutions by an invariant subset X (a global attractor) of the phase space H to which all orbits converge as $t \to \infty$. For dissipative systems, there are many results (see, e.g., Ghidaglia [65, 66], Temam [147], and Goubet [69, 70]) establishing the existence of a compact set that satisfies the above properties. In some cases, the global attractor is a "thin" set; for example, it may be a finite dimensional set, although the phase space is infinite dimensional; Ghidaglia [65].

To describe the problem analytically, we need some definitions from Temam's book on the topic [147]. Let $U(t) : H \to H$ be the globally defined data-to-solution map associated with a dissipative dynamical system on a phase space H.

Definition 5.12 A set $A \subset H$ is called invariant under the flow if

$$U(t)A = A, \quad \text{for all } t > 0.$$

Definition 5.13 An attractor is a set $A \subset H$ that is invariant under the flow and possesses an open neighborhood O such that, for every $g \in O$

$$d(U(t)g, A) \to 0 \text{ as } t \to \infty.$$

Here, the distance is understood to be the distance of a point to the set

$$d(x, Y) = \inf_{y \in Y} d(x, y).$$

We say that A attracts the points of O, and we call the largest open such set O the basin of attraction.

Definition 5.14 We say that $A \subset H$ is a global attractor for the semigroup $\{U(t)\}_{t\geq 0}$ if A is a compact attractor whose basin of attraction is the whole phase space H.

To state a general theorem for the existence of a global attractor, we need one more definition:

Definition 5.15 Let B be a bounded subset of H. We say that B is an absorbing set if for any bounded $S \subset H$ there exists $T = T(S)$ such that for all $t \geq T$, $U(t)S \subset B$.

It is not hard to see that the existence of a global attractor A for a semigroup $\{U(t)\}_{t\geq 0}$ implies the existence of an absorbing set. For the converse, we cite the following theorem from Temam [147], which gives a general criterion for the existence of a global attractor:

Theorem 5.16 *Assume that H is a metric space and that $\{U(t)\}_{t\geq 0}$ is a continuous semigroup from H to itself. Also assume that there exists an absorbing set B. If the semigroup $\{U(t)\}_{t\geq 0}$ is asymptotically compact, i.e. for every bounded sequence x_k in H and every sequence $t_k \to \infty$, $\{U(t_k)x_k\}_k$ is relatively compact in H, then the omega limit set*

$$\omega(B) = \bigcap_{s\geq 0} \overline{\bigcup_{t\geq s} U(t)B}$$

is the unique global attractor. Here the closure is taken on H.

In this section, we present a method for establishing the existence and uniqueness of global attractors using the smoothing estimates we have in Section 4.1. We discuss this method for the forced and weakly damped KdV equation on the torus

$$\begin{cases} u_t + u_{xxx} + \gamma u + uu_x = f, & t \in \mathbb{R}^+, \; x \in \mathbb{T}, \\ u(0,x) = g(x) \in \dot{L}^2(\mathbb{T}), \end{cases} \tag{5.6}$$

where $\gamma > 0$ and $f \in \dot{L}^2(\mathbb{T})$. Here

$$\dot{L}^2(\mathbb{T}) := \left\{ g \in L^2(\mathbb{T}) : \int_{\mathbb{T}} g(x)dx = 0 \right\}.$$

We also assume that u and f are real valued and that f is time independent. Recall that the solution u remains mean-zero at all times. For the forced and weakly damped KdV equation, the conservation of energy does not hold. Nevertheless, by (3.33) the energy remains bounded for positive times

$$\|u(t)\|_{L^2} \le e^{-\gamma t}\|g\|_{L^2} + \frac{\|f\|_{L^2}}{\gamma}(1 - e^{-\gamma t}). \tag{5.7}$$

Thus, for $t > T = T(\gamma, \|g\|_{L^2}, \|f\|_{L^2})$, we have $\|u(t)\|_{L^2} < 2\|f\|_{L^2}/\gamma$, and hence

$$B = \left\{g \in \dot{L}^2(\mathbb{T}) : \|g\|_{L^2} \le \frac{2}{\gamma}\|f\|_{L^2}\right\}$$

is an absorbing set. The bound (5.7) also implies that the solution operator maps the set $\{g : \|g\|_{L^2} \le \|f\|_{L^2}/\gamma\}$ into itself. Recall that the wellposedness theory was established in Section 3.3. Moreover, the bound (5.7) implies the global wellposedness for (5.6) in L^2. For the global wellposedness theory below L^2; see Tsugawa [150] and the references therein.

In the case of the forced and weakly damped KdV, Ghidaglia in [65] established the existence of a global attractor in H^2 for the weak topology. Moreover, the attractor has a finite Hausdorff dimension in H^1. The result can then be upgraded to a result in the strong topology by an argument of Ball [3]; also see Ghidaglia [66]. There are usually two steps in proving such a result. The existence of absorbing sets is derived by establishing energy inequalities coming from the equation, as in (5.7). To prove the asymptotic compactness of the semigroup, one relies again on energy inequalities and the fact that the semigoup is a continuous mapping for the weak topology of H^2. Notice that the continuity of $U(t)$ in H^2 does not imply the weak continuity, which is defined by

$$g_n \overset{w}{\rightarrow} g \text{ in } H^2 \implies u_n(t) \overset{w}{\rightarrow} u(t) \text{ in } H^2.$$

To this end, one uses the fact that the mapping $U(t)$ is continuous with respect to the H^1 norm on bounded subsets of H^2, for details; see Ghidaglia [65]. Ball's argument uses L^2 energy identities to upgrade the asymptotic compactness from the weak to the strong topology.

In [69], Goubet proved the existence of a global attractor on $\dot{L}^2$ using the $X^{s,b}$ theory of Bourgain presented in Section 3.3. Concerning its regularity, he proved that the global attractor is a compact subset of H^3. This was achieved by splitting the solution into two parts, high and low frequencies. The low frequencies are regular and thus in H^3, while the high frequencies decay to zero in L^2 as time goes to infinity. For an alternative way of obtaining the existence and uniqueness of the global attractor in $\dot{L}^2$, see [57] and Theorem 5.17 below. The existence of a global attractor below L^2 was established by Tsugawa in [150]. The difficulty there lies in the fact that there is no conservation law for the KdV equation below L^2. Tsugawa bypasses this problem by using the I–method presented in Section 4.3. In addition, he proves that the global attractor below L^2 is the same as the one obtained by Goubet [69]. One can lower the Sobolev index further, see Yang [156].

Following [57], we prove below that the hypothesis of Theorem 5.16 can be checked using only Theorem 4.3 leading to:

Theorem 5.17 *Consider the forced and weakly damped KdV equation* (5.6) *on* $\mathbb{T} \times \mathbb{R}^+$ *with* $u(0,x) = g(x) \in \dot{L}^2$. *Then the equation* (5.6) *possesses a global attractor in* $\dot{L}^2$. *Moreover, for any* $s \in (0,1)$ *the global attractor is a compact subset of* H^s, *and is bounded in* H^s *by a constant depending only on* s, γ, *and* $\|f\|_{L^2}$.

Note that this theorem implies that the radius of the attractor set in H^s depends only on s, γ, and $\|f\|_{L^2}$. Moreover, the proof is simpler than the proofs in [65, 69] on the existence of the attractor. This method was also used to obtain the existence of a global attractor in the energy space for the Zakharov system; see [58].

Proof of Theorem 5.17 Let

$$v = \left(\widehat{f}/(ik)^3\right)^{\vee} = \partial_x^{-3} f \in H^3,$$

and $w = u - v$. Then w satisfies

$$\begin{cases} w_t + w_{xxx} + \gamma w + \left(\frac{w^2}{2} + vw\right)_x = F, & x \in \mathbb{T}, \quad t \in \mathbb{R}, \\ w(0,x) = g(x) - v(x) \in \dot{L}^2(\mathbb{T}), \end{cases} \tag{5.8}$$

where $F = -\gamma v - vv_x \in H^2$. By applying Theorem 4.3 to (5.8) with time independent $V = v \in H^3$ and $f = F = -\gamma v - vv_x \in H^2$, we obtain:

Theorem 5.18 *Fix* $s \in (0,1)$. *Consider the forced and weakly damped KdV equation* (5.6) *on* $\mathbb{T} \times \mathbb{R}$ *with* $u(x,0) = g(x) \in \dot{L}^2$. *Then*

$$u(t) - e^{tL_\gamma} g \in C_t^0 H_x^s$$

and

$$\left\| u(t) - W_t^\gamma g \right\|_{H^s} \leq C\left(s, \gamma, \|g\|_{L^2}, \|f\|_{L^2}\right),$$

where $W_t^\gamma = e^{-t(\partial_{xxx} + \gamma)}$.

Theorem 5.18 and (5.7) imply the following:

Corollary 5.19 *Fix* $s \in (0,1)$. *Consider the forced and weakly damped KdV* *equation* (5.6) *on* $\mathbb{T} \times \mathbb{R}^+$ *with* $u(0,x) = g(x) \in \dot{L}^2$. *Then there exists* $T = T(\gamma, \|g\|_{L^2}, \|f\|_{L^2})$ *such that for* $t \geq T$,

$$\left\| u(t) - W^{\gamma}_{t-T} u(T) \right\|_{H^s} \leq C(s, \gamma, \|f\|_{L^2}).$$

This implies that all $\dot{L}^2$ solutions are attracted by a ball in H^s centered at zero of radius depending only on $s, \gamma, \|f\|_{L^2}$. An upper bound for this radius can be calculated explicitly by keeping track of the constants in the proof. Moreover, the description of the dynamics is explicit in the sense that after time T the evolution can be written as a sum of the linear evolution which decays to zero exponentially and a nonlinear evolution contained by the attracting ball. We should note that the attracting ball in H^s that Corollary 5.19 provides is not a global attractor since we do not know whether it is an invariant set.

Now, we check the hypothesis of Theorem 5.16. First, note that the existence of an absorbing set, B, is immediate from (5.7). Second, we need to verify the asymptotic compactness of the propagator $U(t)$. It suffices to prove that for any sequence $t_k \to \infty$ and for any sequence g_k in B, the sequence $U_{t_k}(g_k)$ has a convergent subsequence in $\dot{L}^2$. To see this, note that by using Theorem 5.18 for $g \in B$, we have

$$U(t)g = W^{\gamma}_t g + N(t)g,$$

where $N(t)g$ is in a ball in H^s with radius depending on $s, \gamma, \|f\|_{L^2}$. Therefore, by Rellich's theorem, the set

$$\{N(t)g : t > 0,\ g \in B\}$$

is precompact in $\dot{L}^2$. Also noting that

$$\|W^{\gamma}_t g\|_{L^2} \lesssim e^{-\gamma t} \to 0, \quad \text{as } t \to \infty$$

uniformly on B, we conclude that $\{U_{t_k} g_k : k \in \mathbb{N}\}$ is precompact in $\dot{L}^2$. Thus, U_t is asymptotically compact. This implies the existence of a global attractor $A \subset \dot{L}^2$ by Theorem 5.16.

We now prove that the attractor set A is a compact subset of H^s for any $s \in (0,1)$. By Rellich's theorem (Exercise 5.5), it suffices to prove that for any $s \in (0,1)$, there exists a closed ball $B_s \subset H^s$ of radius $C(s, \gamma, \|f\|_{L^2})$ such that $A \subset B_s$. By definition

$$A = \bigcap_{\tau \geq 0} \overline{\bigcup_{t \geq \tau} U(t)B} =: \bigcap_{\tau \geq 0} U_\tau.$$

Let B_s be the ball of radius $C(s, \gamma, \|f\|_{L^2})$ (as in Corollary 5.19) centered at zero in H^s. By Corollary 5.19, for $\tau > T$, U_τ is contained in a δ_τ neighborhood N_τ

of B_s in L^2, where $\delta_\tau \to 0$ as τ tends to infinity. Since B_s is a compact subset of L^2, we have

$$A = \bigcap_{\tau \geq 0} U_\tau \subset \bigcap_{\tau \geq 0} N_\tau = B_s.$$

□

5.4.1 The global attractor is trivial for large damping

In this section following Cabral–Rosa [25], we prove that for sufficiently large γ the global attractor of (5.6) consists of a single point, which is the stationary solution of (5.6). Cabral and Rosa also provided numerical results on the properties of the attractor set for small values of the damping parameter establishing the chaotic behavior of the dynamics. See Erdoğan–Marzuola–Newhall–Tzirakis [55] for related results for the Zakharov system.

First, we prove that given f the stationary equation

$$v_{xxx} + \gamma v + v v_x = f \tag{5.9}$$

has a solution $v \in \dot{H}^3$ provided that γ is sufficiently large:

Lemma 5.20 *There is an absolute constant C such that if*

$$\gamma \geq \min\left((2C\|f\|_{\dot{L}^2})^{\frac{2}{3}}, C^2\|f\|_{\dot{L}^2} \right),$$

then (5.9) *has a unique solution in the set*

$$X = \left\{ v \in C^1 : \widehat{v}(0) = 0, \|v_x\|_{L^\infty} \leq C\gamma^{-1/2}\|f\|_{\dot{L}^2} \leq \gamma/2 \right\}.$$

Moreover, the solution belongs to $\dot{H}^3$.

Proof Note that

$$L_\gamma := -\partial_{xxx} - \gamma$$

with domain $\dot{H}^3$ is invertible as an unbounded operator from $\dot{L}^2$ to $\dot{L}^2$, and

$$\widehat{L_\gamma^{-1} g}(k) = \frac{1}{ik^3 - \gamma}\widehat{g}(k).$$

This and the inequality

$$\frac{1}{\left|ik^3 - \gamma\right|} \leq \gamma^{-1},$$

imply that

$$\left\| L_\gamma^{-1} g \right\|_{\dot{L}^2} \leq \gamma^{-1}\|g\|_{\dot{L}^2}.$$

Fix v such that $v_x \in L^\infty$. Note that

$$\left\|v_x L_\gamma^{-1}\right\|_{\dot{L}^2\to\dot{L}^2} \le \|v_x\|_{L^\infty} \left\|L_\gamma^{-1}\right\|_{\dot{L}^2\to\dot{L}^2} \le \gamma^{-1}\|v_x\|_{L^\infty}.$$

Therefore, if $\|v_x\|_{L^\infty} \le \gamma/2$, then $I - v_x L_\gamma^{-1}$ is invertible on $\dot{L}^2$ and

$$\left(I - v_x L_\gamma^{-1}\right)^{-1} = \sum_{j=0}^{\infty} \left(v_x L_\gamma^{-1}\right)^j.$$

We also have

$$\left\|\left(I - v_x L_\gamma^{-1}\right)^{-1}\right\|_{\dot{L}^2\to\dot{L}^2} \le 2,$$

and that $-L_\gamma + v_x$ is invertible

$$\left(-L_\gamma + v_x\right)^{-1} = -L_\gamma^{-1}\left(I - v_x L_\gamma^{-1}\right)^{-1}. \tag{5.10}$$

Let

$$T(v) = \left(-L_\gamma + v_x\right)^{-1} f.$$

Note that fixed points of T solves (5.9). We now prove that T is a contraction on

$$X = \left\{v \in C^1 : \widehat{v}(0) = 0, \|v_x\|_{L^\infty} \le C\gamma^{-1/2}\|f\|_{\dot{L}^2} \le \gamma/2\right\}.$$

First note that for $v \in X$, the invertibility of $-L_\gamma + v_x$ follows since $\|v_x\|_{L^\infty} \le \gamma/2$. To see that T maps X to X, we calculate (using (5.10), Hausdorff–Young and Cauchy–Schwarz inequalities, and Plancherel's theorem)

$$\begin{aligned}\|\partial_x T(v)\|_{L^\infty} \le \|\mathcal{F}(\partial_x T(v))\|_{\ell^1} &= \sum_k \frac{|k|}{|-ik^3+\gamma|}\left|\mathcal{F}[(I - v_x L_\gamma^{-1})^{-1} f](k)\right| \\ &\le \left\|(I - v_x L_\gamma^{-1})^{-1} f\right\|_{\dot{L}^2} \sqrt{\sum_k \frac{k^2}{k^6+\gamma^2}} \\ &\le 2\|f\|_{\dot{L}^2} \sqrt{\sum_k \frac{k^2}{k^6+\gamma^2}}.\end{aligned}$$

This implies that

$$\|\partial_x T(v)\|_{L^\infty} \le C\gamma^{-1/2}\|f\|_{\dot{L}^2}$$

provided that C is large enough. Note that by the resolvent identity

$$A^{-1} - B^{-1} = A^{-1}(B - A)B^{-1},$$

we have

$$Tu - Tv = (-L_\gamma + u_x)^{-1}(v - u)_x(-L_\gamma + v_x)^{-1} f.$$

Therefore, by using the previous calculation twice we obtain

$$\begin{aligned}\|\partial_x(Tu - Tv)\|_{L^\infty} &\leq C\gamma^{-1/2} \left\|(v-u)_x(-L_\gamma + v_x)^{-1} f\right\|_{\dot{L}^2} \\ &\leq C\gamma^{-1/2} \|(v-u)_x\|_{L^\infty} \left\|(-L_\gamma + v_x)^{-1} f\right\|_{\dot{L}^2} \\ &\leq C^2\gamma^{-1} \|f\|_{\dot{L}^2} \|(v-u)_x\|_{L^\infty} .\end{aligned}$$

Thus, T is a contraction since $\gamma > C^2\|f\|_{\dot{L}^2}$, and the statement follows by Banach's fixed point theorem. Finally, the fixed point v belongs to $\dot{H}^3$ by (5.9). □

We now prove that any solution u of (5.6) with initial data in $\dot{L}^2$ converges to the stationary solution v in $\dot{L}^2$. This also implies the uniqueness of v in $\dot{L}^2$. Let $w = u - v$. It is easy to see that w satisfies

$$w_t + w_{xxx} + \gamma w + w w_x + (vw)_x = 0.$$

Multiplying both sides with w and integrating by parts, we obtain (using $\gamma \geq 2\|v_x\|_{L^\infty}$)

$$\frac{d}{dt}\int_{\mathbb{T}} w^2 = \int_{\mathbb{T}} (-2\gamma - v_x) w^2 \leq -\gamma \int_{\mathbb{T}} w^2.$$

Therefore, by Gronwall's inequality, $w \to 0$ in $\dot{L}^2$ as $t \to \infty$.

5.4.2 Bounds on the forced KdV equation

In this section, we discuss the boundedness of higher order Sobolev norms for the forced KdV equation (5.6) in the case $\gamma > 0$. We will also prove that when $\gamma = 0$ Sobolev norms grow at most polynomially.

First, note that the following corollary is an immediate consequence of Theorem 5.18:

Corollary 5.21 *Fix $s \in (0, 1)$. Consider the equation* (5.6) *on $\mathbb{T} \times \mathbb{R}^+$ with $u(0, x) = g(x) \in \dot{H}^s$. Then for $t \geq 0$*

$$\|u(t)\|_{H^s} \leq C(s, \gamma, \|g\|_{H^s}, \|f\|_{L^2}) .$$

Indeed, by Theorem 5.18

$$\left\|u(t) - W_t^\gamma g\right\|_{H^s} \leq C(s, \gamma, \|g\|_{H^s}, \|f\|_{L^2}) .$$

Therefore, for $t \geq 0$

$$\|u(t)\|_{H^s} \leq \left\|W_t^\gamma g\right\|_{H^s} + C(s, \gamma, \|g\|_{L^2}, \|f\|_{L^2}) \leq C(s, \gamma, \|g\|_{H^s}, \|f\|_{L^2}) .$$

For higher order Sobolev norms, we have the following corollary of Theorem 4.3 that can be proved as Theorem 5.18:

Theorem 5.22 *Fix $r \geq 0$ and $s < r+1$. Consider the equation* (5.6) *on* $\mathbb{T}\times\mathbb{R}^+$ *with $u(0,x) = g(x) \in \dot{H}^r$ and $f \in \dot{H}^r$. Assume that there is an a priori bound*

$$\|u(t)\|_{H^r} \leq C(r,\gamma,\|g\|_{H^r},\|f\|_{H^r})$$

for $t \geq 0$. Then

$$u(t) - W_t^{\gamma} g \in C_t^0 H_x^s$$

and

$$\left\|u(t) - W_t^{\gamma} g\right\|_{H^s} \leq C(s,r,\gamma,\|g\|_{H^r},\|f\|_{H^r}).$$

From Theorem 5.22, we obtain the following corollary:

Corollary 5.23 *Fix $r \geq 0$. Consider the equation* (5.6) *on* $\mathbb{T}\times\mathbb{R}^+$ *with $f \in \dot{H}^r$. Then for any $s < r+1$ and $u(0,x) = g(x) \in \dot{H}^s$, we have*

$$\|u(t)\|_{H^s} \leq C(s,\gamma,\|g\|_{H^s},\|f\|_{H^r})$$

for $t \geq 0$. In particular, if f is C^∞, then all H^s norms remain bounded.

Proof To obtain the global bound, we use Theorem 5.22 repeatedly. For $s \in (0,1)$, the statement is Corollary 5.21. For $s < \min(2, r+1)$, use Theorem 5.22 with $r < 1$ and $s < 2$ (note that the first step supplies the needed a priori bound) to obtain the statement as in the proof of Corollary 5.21. Then continue inductively to obtain the corollary. □

We close this section with some remarks on the case $\gamma = 0$. First note that by (3.32), the L^2 norm grows at most linearly at infinity

$$\|u\|_{L^2} \lesssim (1+|t|)\|f\|_{L^2}.$$

By an iteration argument similar to the one in Corollary 5.23, one can prove that the H^s norms grow at most polynomially provided that f is sufficiently smooth ($f \in H^s$ suffices). This involves proving a variant of Theorem 5.22 in the case $\gamma = 0$, which also follows from Theorem 4.3.

Exercises

5.1 Show that the focusing fractional NLS equation (5.1) is globally wellposed in $H^{\alpha}(\mathbb{T})$.

5.2 Prove the inequality (5.3).
Hint: Repeat the proof of Proposition 5.2 for $R(u)$ by considering the contribution of the term $k = j = 0$ in M_n.

5.3 Describe the resonant set of the half-wave equation

$$\begin{cases} iu_t + Du = |u|^2 u, & x \in \mathbb{T},\ t \in \mathbb{R}, \\ u(0,x) = g(x). \end{cases}$$

5.4 Prove Corollary 5.9. The claim $u \in C^\alpha$ follows from Theorem 2.16 and Theorem 4.1 since $H^{1-} \subset C^\alpha$ for $\alpha < \frac{1}{2}$. The second part follows from the first part by using the arguments in the proof of Theorem 2.16.

5.5 (Rellich's Theorem) Prove by passing to the Fourier side that $H^s(\mathbb{T})$ is compact in $H^r(\mathbb{T})$ for any $s > r$.

5.6 Complete details of the proof of Corollary 5.23.

5.7 In this exercise, we construct a stationary solution of the dissipative Zakharov system with forcing

$$\begin{cases} iu_t + u_{xx} + i\gamma u = nu + f, & x \in \mathbb{T}, \quad t \in [0,\infty), \\ n_{tt} - n_{xx} + \gamma n_t = (|u|^2)_{xx}, \\ u(0,x) = u_0(x) \in H^1(\mathbb{T}), \\ n(0,x) = n_0(x) \in L^2(\mathbb{T}), \quad n_t(0,x) = n_1(x) \in H^{-1}(\mathbb{T}), \quad f \in H^1(\mathbb{T}). \end{cases}$$

Here $\gamma > 0$, n is real, and n and n_t are mean-zero.

(a) Show that a stationary solution (v, m) of the system above satisfies

$$m = -|v|^2 + \frac{1}{2\pi}\|v\|_{L^2}^2,$$

and v solves

$$\left[\frac{\partial^2}{\partial x^2} + i\gamma - \frac{1}{2\pi}\|v\|_{L^2}^2 + |v|^2\right] v = f, \quad x \in \mathbb{T}. \tag{5.11}$$

(b) By multiplying (5.11) with $\overline{v}$ and integrating on $\mathbb{T}$, prove the following a priori estimates

$$\|v\|_{L^2} \leq \frac{1}{\gamma}\|f\|_{L^2}, \tag{5.12}$$

$$\|v_x\|_{L^2} \leq C \max\left(\gamma^{-3}\|f\|_{L^2}^3, \gamma^{-2}\|f\|_{L^2}^2, \gamma^{-1/2}\|f\|_{L^2}\right). \tag{5.13}$$

(c) Prove that for given $f \in H^1$, if γ is sufficiently large, then (5.11) has a unique solution in the ball

$$\left\{v : \|v\|_{H^1} \leq \frac{2}{\gamma}\|f\|_{H^1}\right\}$$

by arguments similar to the proof of Lemma 5.20.

References

[1] A. Babin, A. A. Ilyin, and E. S. Titi, On the regularization mechanism for the periodic Korteweg–de Vries equation, *Comm. Pure Appl. Math.* **64** (2011), no. 5, 591–648.

[2] A. Babin, A. Mahalov, and B. Nicolaenko, Regularity and integrability of 3D Euler and Navier–Stokes equations for rotating fluids, *Asymptot. Anal.* **15** (1997), no.2, 103–150.

[3] J. M. Ball, Global attractors for damped semilinear wave equations, *Discrete Contin. Dyn. Syst.* **10** (2004), no. 1–2, 31–52.

[4] M. Beals, Self-spreading and strength of singularities of solutions to semilinear wave equations, *Ann. of Math.* **118** (1983), 187–214.

[5] H. Berestycki and P.-L. Lions, Existence d'ondes solitaires dans des problemes nonlinéaires du type Klein–Gordon, *C. R. Acad. Sci. Paris Sér. A-B* **288** (1979), no. 7, A395–A398.

[6] M. V. Berry, Quantum fractals in boxes, *J. Phys. A: Math. Gen.* **29** (1996), 6617–6629.

[7] M. V. Berry and S. Klein, Integer, fractional and fractal Talbot effects, *J. Mod. Optics* **43** (1996), 2139–2164.

[8] M. V. Berry and Z. V. Lewis, On the Weierstrass–Mandelbrot fractal function, *Proc. Roy. Soc. London A* **370** (1980), 459–484.

[9] M. V. Berry, I. Marzoli, and W. Schleich, Quantum carpets, carpets of light, *Physics World* **14** (2001), no.6, 39–44.

[10] H. Biagioni and F. Linares, Ill-posedness for the Derivative Schrödinger and Generalized Benjamin-Ono equations, *Trans. Amer. Math. Soc.* **353** (2001), 3649–3659.

[11] B. Birnir, C. E. Kenig, G. Ponce, N. Svanstedt, and L. Vega, On the ill-posedness of the IVP for the generalized Korteweg–de Vries and nonlinear Schrödinger equations, *J. London Math. Soc.* **53** (1996), pp. 551–559.

[12] E. Bombieri and J. Pila, The number of integral points on arcs and ovals, *Duke Math. J.* **59** (1989), 337–357.

[13] J. L. Bona and R. Smith, The initial-value problem for the Korteweg–de Vries equation, *Philos. Trans. Roy. Soc. London Ser. A* **278** (1975), no. 1287, 555–601.

[14] J. L. Bona and N. Tzvetkov, Sharp wellposedness results for the BBM equation, *Discrete Contin. Dyn. Syst.* **23** (2009), no.4, 1241–1252.

[15] J. Bourgain, Fourier transform restriction phenomena for certain lattice subsets and applications to nonlinear evolution equations. Part I: Schrödinger equations, *GAFA* **3** (1993), 209–262.

[16] J. Bourgain, Fourier transform restriction phenomena for certain lattice subsets and applications to nonlinear evolution equations. Part II: The KdV equation, *GAFA* **3** (1993), 209–262.

[17] J. Bourgain, On the growth in time of higher Sobolev norms of smooth solutions of Hamiltonian PDE, *Internat. Math. Res. Notices* (1996), no. 6, 277–304.

[18] J. Bourgain, Periodic Korteweg–de Vries equation with measures as initial data, *Sel. Math. New ser.* **3** (1997), 115–159.

[19] J. Bourgain, Refinement of Strichartz' inequality and applications to 2D–NLS with critical nonlinearity, *Internat. Math. Res. Notices* (1998), no. 5, 253–283.

[20] J. Bourgain, *Global Solutions of Nonlinear Schrödinger Equations*, AMS Colloquium Publications **46**. American Mathematical Society, Providence, RI, 1999.

[21] J. Bourgain, A remark on normal forms and the "I-method" for periodic NLS, *J. Anal. Math.* **94** (2004), 125–157.

[22] J. Bourgain and C. Demeter, The proof of the ℓ^2 decoupling conjecture, preprint 2014, http://arxiv.org/abs/1403.5335.

[23] N. Burq, P. Gérard, and N. Tzvetkov, An instability property of the nonlinear Schrödinger equation on S^d, *Math. Res. Lett.* **9** (2002), 323–335.

[24] N. Burq, P. Gérard, and N. Tzvetkov, Global solutions for the nonlinear Schrödinger equation on three-dimensional compact manifolds, *Mathematical Aspects of Nonlinear Dispersive Equations*, 111–129, Ann. of Math. Stud., **163**, Princeton Univ. Press, Princeton, NJ, 2007.

[25] M. Cabral and R. Rosa, Chaos for a damped and forced KdV equation, *Physica D* **192** (2004), 265–278.

[26] L. Carleson, Some analytical problems related to statistical mechanics, *Euclidean Harmonic Analysis*, 5–45, Lecture Notes in Math. **779** Springer-Verlag, Berlin and New York, 1979.

[27] F. Catoire and W.-M. Wang, Bounds on Sobolev norms for the defocusing nonlinear Schrödinger equation on general flat tori, *Commun. Pure Appl. Anal.* **9** (2010), no. 2, 483–491.

[28] T. Cazenave, *An Introduction to Nonlinear Schrödinger Equations*, Textos de Métodos Matemáticos, **22**, Instituto de Matemática, Universidade Federal do Rio de Janeiro, Rio de Janeiro, 1989.

[29] T. Cazenave, *Blow Up and Scattering in the Nonlinear Schrödinger Equation*, Textos de Métodos Matemáticos, **30**, Instituto de Matemática, Universidade Federal do Rio de Janeiro, Rio de Janeiro, 1994.

[30] T. Cazenave, *Semilinear Schrödinger Equations,* Courant Lecture Notes in Mathematics **10**. New York University, Courant Institute of Mathematical Sciences, NY; American Mathematical Society, Providence, RI, 2003.

[31] T. Cazenave and F. B. Weissler, Some remarks on the nonlinear Schrödinger equation in the critical case. *Nonlinear Semigroups, Partial Differential Equations and Attractors* (Washington, DC, 1987), 18–29, Lecture Notes in Math., **1394**, Springer, Berlin, 1989.

[32] N. Chang, J. Shatah, and K. Uhlenbeck, Schrödinger maps, *Comm. Pure Appl. Math.* **53** (2000), 590–602.

[33] G. Chen and P. J. Olver, Numerical simulation of nonlinear dispersive quantization, *Discrete Contin. Dyn. Syst.* **34** (2014), no. 3, 991–1008.

[34] G. Chen and P. J. Olver, Dispersion of discontinuous periodic waves, *Proc. R. Soc. Lond. Ser. A Math. Phys. Eng. Sci.* **469** (2013), no. 2149, 20120407, 21 pp.

[35] Y. Cho, G. Hwang, S. Kwon, and S. Lee, well-posedness and ill-posedness for the cubic fractional Schrödinger equations, *Discrete Contin. Dyn. Syst.* **35** (2015), no. 7, 2863–2880.

[36] V. Chousionis, M.B. Erdoğan, and N. Tzirakis, Fractal solutions of linear and nonlinear dispersive partial differential equations, *Proc. Lond. Math. Soc.* **110** (2015), no. 3, 543–564.

[37] M. Christ, Power series solution of a nonlinear Schrödinger equation in *Mathematical Aspects of Nonlinear Dispersive Equations*, 131–155, Ann. of Math. Stud., **163**, Princeton Univ. Press, Princeton, NJ, 2007.

[38] M. Christ, J. Colliander, and T. Tao, Asymptotics, frequency modulation, low regularity ill-posedness for canonical defocusing equations, *Amer. J. Math.* **125** (2003), 1235–1293.

[39] M. Christ, J. Colliander, and T. Tao, Instability of the periodic nonlinear Schrödinger equation, arxiv:math/0311227v1.

[40] J. Chung, Z. Guo, S. Kwon, and T. Oh, Normal form approach to global well-posedness of the quadratic derivative Schrödinger equation on the circle, Preprint 2015, http://arxiv.org/abs/1509.08139.

[41] J. Colliander, M. Keel, G. Staffilani, H. Takaoka, and T. Tao, Global well-posedness for Schrödinger equation with derivative, *SIAM J. Math. Anal.* **33** (2001), 649–669.

[42] J. Colliander, M. Keel, G. Staffilani, H. Takaoka, T. Tao, Almost conservation laws and global rough solutions to a nonlinear Schrödinger equation, *Math. Research Letters*, **9** (2002), 659–682.

[43] J. Colliander, M. Keel, G. Staffilani, H. Takaoka, and T. Tao, Sharp global well-posedness for KdV and modified KdV on $\mathbb{R}$ and $\mathbb{T}$, *J. Amer. Math. Soc.* **16** (2003), 705–749.

[44] J. Colliander, M. Keel, G. Staffilani, H. Takaoka, and T. Tao, Global existence and scattering for rough solutions of a nonlinear Schrödinger equation on $\mathbb{R}^3$, *Comm. Pure Appl. Math.* **57** (2004), no. 8, 987–1014.

[45] J. Colliander, S. Kwon, and T. Oh, A remark on normal forms and the "upside-down" I-method for periodic NLS: growth of higher Sobolev norms, *J. Anal. Math.* **118** (2012), no. 1, 55–82.

[46] J. Colliander, G. Staffilani, and H. Takaoka, Global wellposedness for KdV below L^2, *Math. Res. Lett.* **6** (1999), no. 5–6, 755–778.

[47] E. Compaan, Smoothing and global attractors for the Majda–Biello system on the torus, *Differential Integral Equations* **29** (2016), no. 3–4, 269–308.

[48] B. E. Dahlberg and C. E. Kenig, A note on the almost everywhere behavior of solutions to the Schrödinger equation, *Harmonic Analysis* (Minneapolis, Minn., 1981), 205–209, Lecture Notes in Math. **908**, Springer, Berlin-New York, 1982.

[49] A. Deliu and B. Jawerth, Geometrical dimension versus smoothness, *Constr. Approx.* **8** (1992), 211–222.

[50] S. Demirbaş, Local well-posedness for 2-D Schrödinger equation on irrational tori and bounds on Sobolev norms, Preprint 2013, to appear in *Communications in Pure and Applied Analysis,* http://arxiv.org/abs/1307.0051.

[51] S. Demirbaş, M. B. Erdoğan, and N. Tzirakis, Existence and uniqueness theory for the fractional Schrödinger equation on the torus, http://arxiv.org/abs/1312.5249.

[52] B. Dodson, Global well-posedness and scattering for the defocusing, L^2 critical, nonlinear Schrödinger equation when $d = 1$, http://arxiv.org/abs/1010.0040.

[53] B. Dodson, Global well-posedness and scattering for the mass critical nonlinear Schrödinger equation with mass below the mass of the ground state, preprint 2011, http://arxiv.org/abs/1104.1114.

[54] P. F. Embid and A. J. Majda, Averaging over fast gravity waves for geophysical flows with arbitrary potential vorticity, *Comm. Partial Diff. Eqs.* **21** (1996), 619–658.

[55] M. B. Erdoğan, J. Marzuola, K. Newhall, and N. Tzirakis, Global Attractors for Forced and Weakly Damped Zakharov System on the Torus, *SIAM J. Appl. Dyn. Syst.* **14** (2015), 1978–1990.

[56] M. B. Erdoğan and N. Tzirakis, Global smoothing for the periodic KdV evolution, *Internat. Math. Res. Notices* (2013), 4589–4614.

[57] M. B. Erdoğan and N. Tzirakis, Long time dynamics for forced and weakly damped KdV on the torus, *Commun. Pure Appl. Anal.* **12** (2013), 2669–2684.

[58] M. B. Erdoğan and N. Tzirakis, Smoothing and global attractors for the Zakharov system on the torus, *Analysis & PDE* **6** (2013), no. 3, 723–750.

[59] M. B. Erdoğan and N. Tzirakis, Talbot effect for the cubic nonlinear Schrödinger equation on the torus, *Math. Res. Lett.* **20** (2013), 1081–1090.

[60] M. B. Erdoğan and N. Tzirakis, Regularity properties of the cubic nonlinear Schrödinger equation on the half line, http://arxiv.org/abs/1509.03546.

[61] G. B. Folland, *Real analysis. Modern Techniques and Their Applications*, Pure and Applied Mathematics, John Wiley & Sons, Inc., NY, 1984.

[62] G. Fonseca, F. Linares, and G. Ponce, Global well-posedness for the modified Korteweg–de Vries equation, *Comm. PDE* **24** (1999), 683–705.

[63] G. Fonseca, F. Linares, and G. Ponce, Global existence for the critical generalized KdV equation, *Proc. Amer. Math. Soc.* **131** (2003), 1847–1855.

[64] P. Germain, N. Masmudi, and J. Shatah, Global solutions for the gravity water waves equation in dimension 3, *C. R. Math. Acad. Sci. Paris* **347** (2009), 897–902.

[65] J. M. Ghidaglia, Weakly damped forced Korteweg–de Vries equations behave as a finite dimensional dynamical system in the long time, *J. Diff. Eqs.* **74** (1988), 369–390.

[66] J. M. Ghidaglia, A note on the strong convergence towards attractors of damped forced KdV equations, *J. Diff. Eqs.* **110** (1994), 356–359.

[67] J. Ginibre and G. Velo, The global Cauchy problem for the nonlinear Schrödinger equation revisited, *Ann. Inst. H. Poincaré Annal. Non linéaire* **2** (1985), 309–327.

[68] R. T. Glassey, On the blowing up of solutions to the Cauchy problem for nonlinear Schrödinger equations, *J. Math. Phys.* **18** (1977), no. 9, 1794–1797.

[69] O. Goubet, Asymptotic smoothing effect for weakly damped forced Korteweg–de Vries equations, *Discrete Contin. Dyn. Syst.* **6** (2000), 625–644.

[70] O. Goubet, Asymptotic smoothing effect for a weakly damped nonlinear Schrödinger equation in $\mathbb{T}^2$, *J. Diff. Eqs.* **165** (2000), no. 1, 96–122.

[71] D. J. Griffiths, *Introduction to Quantum Mechanics*, Prentice Hall, 1995.

[72] Z. Guo, Global well-posedness of Korteweg–de Vries equation in $H^{3/4}(\mathbb{R})$, *J. Math. Pures Appl.* **91** (2009), no. 6, 583–597.

[73] Z. Guo, Local well-posedness for dispersion generalized Benjamin–Ono equations in Sobolev spaces, *J. Diff. Eqs.* **252** (2012), 2053–2084.

[74] N. J. Hitchin, G. B. Segal, and R. S. Ward, *Integrable Systems: Twistors, Loop Groups, and Riemann Surfaces,* Oxford University Press, 1999.

[75] Y. Hu and X. Li, Discrete Fourier restriction associated with KdV equations, *Anal. PDE* **6** (2013), no. 4, 859–892.

[76] T. W. Hungerford, *Algebra*, Holt, Rinehart & Winston, Inc., New York–Montreal, Que.–London, 1974.

[77] A. D. Ionescu and C. E. Kenig, Global well-posedness of the Benjamin–Ono equation in low-regularity spaces, *J. Amer. Math. Soc.* **20** (2007), 753-798.

[78] L. Kapitanski and I. Rodnianski, Does a quantum particle knows the time?, in: *Emerging Applications of Number Theory,* D. Hejhal, J. Friedman, M. C. Gutzwiller and A. M. Odlyzko, eds., IMA Volumes in Mathematics and Its Applications, vol. **109**, Springer Verlag, New York, 1999, pp. 355–371.

[79] T. Kappeler, B. Schaad, and P. Topalov, Qualitative features of periodic solutions of KdV, *Comm. PDE* **38** (2013), 1626–1673.

[80] T. Kappeler, B. Schaad, and P. Topalov, Scattering-like phenomena of the periodic defocusing NLS equation, http://arxiv.org/abs/1505.07394.

[81] T. Kappeler and P. Topalov, Global wellposedness of KdV in $H^{-1}(\mathbb{T},\mathbb{R})$, *Duke Math. J.* **135**, no. 2 (2006), 327–360.

[82] T. Kato, On the Korteweg–de Vries equation, *Manuscripta Math.* **28** (1979), 89–99.

[83] T. Kato and G. Ponce, Commutator estimates and the Euler and Navier–Stokes equations, *Commun. Pure Appl. Math.* **41** (1988), 891–907.

[84] Y. Katznelson, *An Introduction to Harmonic Analysis.* Third edition, Cambridge Mathematical Library, Cambridge University Press, Cambridge, 2004.

[85] M. Keel and T. Tao, Endpoint Strichartz estimates, *Amer. J. Math.* **120** (1998), no. 5, 955–980.

[86] M. Keel and T. Tao, Local and global well-posedness of wave maps on R^{1+1} for rough data, *Internat. Math. Res. Notices* 1998, no. 21, 1117–1156.

[87] C. E. Kenig, The Cauchy problem for the quasilinear Schrödinger equation, http://arxiv.org/abs/1309.3291.

[88] C. E. Kenig, G. Ponce, and L. Vega, Well-posedness and scattering results for the generalized Korteweg–de Vries equation via the contraction principle, *Comm. Pure Appl. Math.* **46** (1993), 527–620.

[89] C. E. Kenig, G. Ponce, and L. Vega, Small solutions to nonlinear Schrödinger equations, *Ann. Inst. H. Poincaré Anal. Non Linéaire* **10** (1993), no. 3, 255–288.

[90] C. E. Kenig, G. Ponce, and L. Vega, On the Zakharov and Zakharov–Schulman systems, *J. Funct. Anal.* **127** (1995), no. 1, 204–234.

[91] C. E. Kenig, G. Ponce, and L. Vega, A bilinear estimate with applications to the KdV equation, *J. Amer. Math.* Soc. **9** (1996), no. 2, 573–603.
[92] C. E. Kenig, G. Ponce, and L. Vega, On the ill-posedness of some canonical dispersive equations, *Duke Math. J.* **106** (2001), 617–633.
[93] S. Keraani and A. Vargas, A smoothing property for the L^2-critical NLS equations and an application to blow up theory, *Ann. Inst. H. Poincaré Annal. Non linéaire* **26** (2009), 745–762.
[94] A. Y. Khinchin, *Continued Fractions*, Translated from the 3rd Russian edition of 1961, The University of Chicago Press, 1964.
[95] N. Kishimoto, Remark on the periodic mass critical nonlinear Schrödinger equation, *Proc. Amer. Math. Soc.* **142** (2014), 2649–2660.
[96] S. Klainerman and M. Machedon, space-time estimates for null forms and a local existence theorem, *Comm. Pure Appl. Math.* **46** (1993), 1221–1268.
[97] H. Koch and N. Tzvetkov, Nonlinear wave interactions for the Benjamin-Ono equation, *Internat. Math. Res. Notices* 2005, no. 30, 1833–1847.
[98] L. Kuipers and H. Niederreiter, *Uniform Distribution of Sequences*, Pure and Applied Mathematics. Wiley-Interscience, New York-London-Sydney, 1974.
[99] S. B. Kuksin, *Analysis of Hamiltonian PDEs* , Oxford University Press, 2000.
[100] S. Kwon and T. Oh, On unconditional well-posedness of modified KdV, *Internat. Math. Res. Notices* 2012, no. 15, 3509–3534.
[101] M. K. Kwong, Uniqueness of positive solutions of $\Delta u - u + up = 0$ in $\mathbb{R}^n$, *Arch. Rational Mech. Anal.* **105** (1989), no. 3, 243–266.
[102] P. Lévy, *Théorie de l'Addition des Variables Aléatoires,* Paris, 1937.
[103] Y. Li, Y. Wu, and G. Xu, Global well-posedness for the mass-critical nonlinear Schrödinger equation on $\mathbb{T}$, *J. Diff. Eqs.* **250** (2011), no. 6, 2715–2736.
[104] F. Linares, D. Pilod, and J.-C. Saut, Dispersive perturbations of Burgers and hyperbolic equations I: Local theory, *SIAM J. Math. Anal.* **46** (2014), no. 2, 1505-1537.
[105] F. Linares and G. Ponce, *Introduction to Nonlinear Dispersive Equations,* UTX, Springer–Verlag, 2009.
[106] L. Molinet, Global well-posedness in the energy space for the Benjamin–Ono equation on the circle, *Math. Ann.* **337** (2007), 353–383.
[107] L. Molinet, On ill-posedness for the one-dimensional periodic cubic Schrödinger equation, *Math. Res. Lett.* **16** (2009), 111-120.
[108] L. Molinet, Sharp ill-posedness result for the periodic Benjamin–Ono equation, *J. Funct. Anal.* **257**, (2009), 3488-3516.
[109] L. Molinet, A note on ill-posedness for the KdV equation, *Differential Integral Equations* **24** (2011), 759–765.
[110] L. Molinet, Sharp ill-posedness results for the KdV and mKdV equations on the torus, *Adv. Math.* **230** (2012), 1895–1930.
[111] L. Molinet, Global well-posedness in L^2 for the periodic Benjamin-Ono equation, *Amer. J. Math.* **130** (2008), no. 3, 635-683.
[112] L. Molinet, J. C. Saut, and N. Tzvetkov, Ill-posedness issues for the Benjamin–Ono and related equations, *Siam J. Math. Anal.* **33** (2001), 982–988.
[113] H. L. Montgomery, *Ten Lectures on the Interface Between Analytic Number Theory and Harmonic Analysis*, CBMS Regional conference series in Mathematics Volume **84**, 1990.

[114] S J. Montgomery-Smith, Time decay for the bounded mean oscillation of solutions of the Schrödinger and wave equation, *Duke Math. J.* **19** (1998), 393–408.

[115] A. Moyua and L. Vega, Bounds for the maximal function associated to periodic solutions of one-dimensional dispersive equations, *Bull. Lond. Math. Soc.* **40** (2008), no. 1, 117–128.

[116] C. Muscalu and W. Schlag, *Classical and Multilinear Harmonic Analysis.* Vol. I. Cambridge Studies in Advanced Mathematics, **137**, Cambridge University Press, Cambridge, 2013.

[117] L. Nirenberg, On elliptic partial differential equations, *Ann. Scuola Norm. Sup. Pisa* **13** (1959), 115–162.

[118] T. Ogawa and Y. Tsutsumi, Blow-up of H^1 solutions for the one-dimensional nonlinear Schrödinger equation with critical power nonlinearity, *Proc. Amer. Math. Soc.* **111** (1991), no. 2, 487–496.

[119] S. Oh, Resonant phase–shift and global smoothing of the periodic Korteweg–de Vries equation in low regularity settings, *Adv. Diff. Eqs.* **18** (2013), 633–662.

[120] P. J. Olver, Dispersive quantization, *Amer. Math. Monthly* **117** (2010), no. 7, 599–610.

[121] K. I. Oskolkov, A class of I. M. Vinogradov's series and its applications in harmonic analysis, in: *Progress in Approximation Theory* (Tampa, FL, 1990), Springer Ser. Comput. Math. **19**, Springer, New York, 1992, pp. 353–402.

[122] H. Pecher, Global well-posedness below energy space for the 1-dimensional Zakharov system, *Internat. Math. Res. Notices* 2001, no. 19, 1027–1056.

[123] J. Pöschel and E. Trubowitz, *Inverse Spectral Theory*, Pure Appl. Math. **130**, Academic Press, Boston, 1987.

[124] L. Rayleigh, On copying diffraction-gratings, and on some phenomena connected therewith, *Philos. Mag.* **11** (1881), 196–205.

[125] I. Rodnianski, Fractal solutions of the Schrödinger equation, *Contemp. Math.* **255** (2000), 181–187.

[126] J. C. Saut and R. Temam, Remarks on the Korteweg–de Vries equation, *Israel J. Math.* **24** (1976), 78–87.

[127] J. Shatah, Normal forms and quadratic nonlinear Klein–Gordon equations, *Comm. Pure Appl. Math.* **38** (1985), 685–696.

[128] V. Sohinger, Bounds on the growth of high Sobolev norms of solutions to nonlinear Schrödinger equations on S^1, *Differential and Integral Equations* **24** (2011), no. 7–8, 653–718.

[129] V. Sohinger, Bounds on the growth of high Sobolev norms of solutions to nonlinear Schrödinger equations on $\mathbb{R}$, *Indiana Math. J.* **60** (2011), no. 5, 1487–1516.

[130] G. Staffilani, On the growth of high Sobolev norms of solutions for KdV and Schrödinger equations, *Duke Math. J.* **86** (1997), no. 1, 109–142.

[131] A. Stefanov, Strichartz estimates for the Schrödinger equation with radial data, *Proc. Amer. Math. Soc.* **129** (2001), 1395–1401.

[132] E. M. Stein, Interpolation of linear operators, *Trans. Amer. Math. Soc.* **83** (1956), 482–492.

[133] E. M. Stein, *Harmonic Analysis: Real-Variable Methods, Orthogonality, and Oscillatory Integrals,* Princeton Mathematical Series, **43**. Monographs in Harmonic Analysis, III. Princeton University Press, Princeton, NJ, 1993.

[134] E. M. Stein and R. Shakarchi, *Functional Analysis. Introduction to Further Topics in Analysis.* Princeton Lectures in Analysis, **4**. Princeton University Press, Princeton, NJ, 2011.

[135] E. M. Stein and G. Weiss, *Introduction to Fourier Analysis on Euclidean Spaces,* Princeton Mathematical Series, **32**. Princeton University Press, Princeton, NJ, 1971.

[136] R. S. Strichartz, Restriction of Fourier transform to quadratic surfaces and decay of solutions of wave equations, *Duke Math. J.* **44** (1977), 705–714.

[137] C. Sulem and P. L. Sulem, *The Nonlinear Schrödinger Equation*, Applied Mathematical Sciences **139**. Springer–Verlag NY 1999.

[138] P.L. Sulem, C. Sulem and C. Bardos, On the continuous limit for a system of classical spins, *Comm. Math. Phys.* **107** (1986), no. 3, 431–454.

[139] H. Takaoka, Y. Tsutsumi, Well-posedness of the Cauchy problem for the modified KdV equation with periodic boundary condition, *Internat. Math. Res. Notices* 2004, no. 56, 3009–3040.

[140] H. F. Talbot, Facts related to optical science, No. IV, *Philo. Mag.* **9** (1836), 401–407.

[141] T. Tao, Spherically averaged endpoint Strichartz estimates for the two–dimensional Schrödinger equation, *Commun. PDE* **25** (2000), 1471–1485.

[142] T. Tao, Global well-posedness of the Benjamin–Ono equation in $H^1(\mathbb{R})$, *J. Hyperbolic Differ. Equ.* **1** (2004), no. 1, 27–49.

[143] T. Tao, *Nonlinear Dispersive Equations: Local and Global Analysis*, CBMS Regional Conference Series in Mathematics **106**. American Mathematical Society, Providence, RI, 2006.

[144] M. Taylor, The Schrödinger equation on spheres, *Pacific J. Math.* **209** (2003), 145–155.

[145] M. Taylor, Tidbits in Harmonic Analysis, Lecture Notes, UNC, 1998.

[146] M. Taylor, *Tools for PDE. Pseudodifferential Operators, Paradifferential Operators, and Layer Potentials,* Mathematical Surveys and Monographs, **81**. American Mathematical Society, Providence, RI, 2000.

[147] R. Temam, *Infinite-Dimensional Dynamical Systems in Mechanics and Phyiscs*, Applied Mathematical Sciences **68**, Springer, 1997.

[148] P. Tomas, A restriction theorem for the Fourier transform, *Bull. Amer. Math. Soc.* **81** (1975), 477–478.

[149] H. Triebel, *Theory of Function Spaces*, Birkhäuser, Basel, 1983.

[150] K. Tsugawa, Existence of the global attractor for weakly damped forced KdV equation on Sobolev spaces of negative index, *Commun. Pure Appl. Anal.* **3** (2004), no. 2, 301–318.

[151] Y. Tsutsumi, L^2 solutions for nonlinear Schrödinger equations and nonlinear groups, *Funkcial. Ekvac.* **30** (1987), 115–125.

[152] N. Tzvetkov, Remark on the local ill-posedness for KdV equation, *C. R. Acad. Sci. Paris* **329** Série I (1999), 1043–1047.

[153] M. Wadati, H. Sanuki, and K. Konno, Relationships among inverse method, Bäcklund transformation and an infinite number of conservation laws, *Progr. Theoret. Phys.* **53** (1975), 419–436.

[154] T. Wolff, *Lectures on Harmonic Analysis,* University Lecture Series, **29**. American Mathematical Society, Providence, RI, 2003.

[155] J. Yang, Nonlinear waves in integrable and nonintegrable systems, *Mathematical Modeling and Computation* **16**, SIAM, Philadelphia, PA, 2010.

[156] X. Yang, Global attractor for the weakly damped forced KdV equation in Sobolev spaces of low regularity, *Nonlinear Differ. Equ. Appl.*, published online.

[157] V. E. Zakharov and A. B. Shabat, Exact theory of two-dimensional self-focusing and one-dimensional self-modulation of waves in nonlinear media, *Soviet Physics JETP* **34** (1972), no. 1, 62–69.

[158] Y. Zhang, J. Wen, S. N. Zhu, and M. Xiao, Nonlinear Talbot effect, *Phys. Rev. Lett.* **104**, 183901 (2010).

[159] A. Zygmund, On Fourier coefficients and transforms of functions of two variables, *Stud. Math.* **50** (1974), 189–201.

Index

Printed in the United States
By Bookmasters